Wilhelm Reich and the
Healing of Atmospheres

By Roberto Maglione, MS

Wilhelm Reich and the
Healing of Atmospheres

By Roberto Maglione, MS

Natural Energy Works
Ashland, Oregon, USA
2007

Translated by Carron Popplewell and Daniela-Sabina Brückner from the Italian work
"Wilhelm Reich e la Modificazione del Clima"

Published by Natural Energy Works, Ashland, Oregon, USA

Distribution to the public and booksellers through:

Ingram Books / Lightning Source

and

Natural Energy Works
PO Box 1148
Ashland, Oregon 97520 USA
E-mail: info@naturalenergyworks.net
www.naturalenergyworks.net

ISBN: 978-0-9802316-6-3   0-9802316-6-3

First English Edition   20111131

Front cover photos reproduced courtesy of the Wilhelm Reich Museum & Trust
Rangeley, Maine, USA   www.wilhelmreichtrust.org

Dedicated to my parents.

Acknowledgments

It is difficult to say why you write a book, particularly on a topic as difficult and complex as the modification and self-regulation of the climate. I don't believe that the basis is a professional or economical motivation. I have seen that very often you write in your spare time and in the breaks from other work. The authors could simply have a message to deliver, which they may or may not be aware of, and they do this in the way most congenial to them.

This last point could be the reason why I published this book which I started to write eight years ago. Spreading some of the least known, least understood, and probably one of the most controversial discoveries of the late Wilhelm Reich: restoring of the natural self-regulating properties of climate, otherwise known as *Cloudbusting*.

It has not been easy to complete this work, due to both the complexity of the topic and the difficulty in finding suitable material. I have to thank many people who have helped me during these years and contributed to producing this book: James DeMeo, Director of the *Orgone Biophysical Research Laboratory* in Ashland (Oregon), for technical information on the *cloudbuster* and information on his research activities during the last 30 years, as well as permission to use material related to his experiences in the field; Bernd Senf, from the *Fachhochschule für Wirtschaft* in Berlin, for information related to the field experiments in Namibia and Eritrea and for the photographs; Richard Blasband, Scientific Director of the *Center for Functional Research* in Tiburon (California) for the photographic material; Charles R. Kelley, founder of the *Radix Institute* in Albuquerque (New Mexico), for the material related to his pioneering experiments; Jonathan Coe for the material related to the operations of Jerome Eden; Trevor J. Constable and George Wuu of the *Etheric Rain Engineering* Pte Ltd. in Singapore for the information on their many activities and the permission to use photographs of their operations; Bernardo Zanini and Carmelodi Doz for their experiences and all the material that they made available to me on their research activities over ten years; Mirko Kulig for supplying material on his experiences; The *Wilhelm Reich Museum* for the permission to use photographs of Reich's activities; Carlo Splendore for the long and continuous discussions on orgone physics, for his precious advice on the scientific content of the book, and for his proofreading of the final manuscript; Carlo Albini for the revisions and critical comments on the final draft of the book; as well as Franco Arborio, Bruna Bortolozzo, and Cristina Vassarotto for their critical considerations on the topic.

Finally, a heartfelt *Thank you* to Ornella for her advice and never-ending patience.

This book is also dedicated to the memory of Luigi Ferrando, a dear friend and teacher, sadly missed.

Table of Contents

Author's Foreword to the English Edition

This English version of my book is being published about two years after the Italian version.

During this period, there has been a growing interest in Wilhelm Reich's research, and especially in *cloudbusting* and other forms of weather modification. The interventions to modify climatic conditions are currently a hot topic, both due to multiple uses of the techniques and current necessity, which is dictated by the environmental problems.

The question I have been asked most frequently since this book came out is why this technology, which was developed more than fifty years ago, and that is presently in both its development and refinement phases, has not yet received the widespread fame it deserves. The technology is trustworthy and can be used to at least temporarily resolve problems such as the presence of DOR, a form of a stagnant atmosphere with high humidity and temperatures that are very common in Northern Italy; smog, forest fires, drought, and for the no less important endeavor of deviating hurricanes away from dry land, out onto the open ocean, and avoiding the disasters that certain areas are frequently subjected to.

It is very difficult to give a direct answer. Maybe people are not yet ready to accept a very simple and economical technology. Or maybe other technologies have precedence for now due to the fact that they are more widespread and more widely used, or because of the enormous investments that were required for their development.

One thing is certain. People need concrete facts. They are tired of facing the great problems that frequently hit our planet with actions that have at best only a slightly improving effect, and at worst make the problem even worse.

There's another thing that I think is indisputable. Reich and his work, especially what he developed in the second part of his life, the orgone biophysics, of which *cloudbusting* is an integral part, it is only little if never studied.

It is carried on only by a few passionate researchers who spend much of their time and money improving and developing it in a serious way. It is essentially important that this new science is not kept isolated and marginalized. It is necessary that it is considered as a potential resource for the future, giving it the importance it deserves. It is also important to start an in-depth study using the current available means, parallel to other sciences, which can only occur through the intervention of governmental departments or private research groups.

The translation of this book has maintained the structure of the Italian version. Corrections have been carried out concerning the inaccuracies that inevitably occur in a book. I have added general information relevant to some important topics such as global weather variations. In addition I have added some parts of the research by Charles R. Kelley, Jerome Eden, James DeMeo and Trevor J. Constable with new photos and information, so as to make the descriptions of the work of these scientists all the more complete.

I hope that this book will be interesting to its readers, but above all I hope it will play an important part in the rational scientific development of the *cloudbusting* technique, as it was intended by the inventor: Wilhelm Reich.

I would also like to emphasize that the information contained in this book must not be, in any way, shape or form, considered and used for field operations, whatever their goal. It is only intended as material to be used with great precaution by very skilled and responsible people either out in the field, under government permission, or in a laboratory research setting, aimed at developing the knowledge and protocols to make *cloudbusting* more and more secure. It was never my intention to spread information that could be used for field applications, either on a small or large scale. I do not, in any way, approve of such actions, and the responsibility lies with the person who might undertake them.

Roberto Maglione
Moncrivello
4 August 2006

Foreword to the English Edition
by James DeMeo

*"Discovery means seeing something everybody has seen
and thinking something nobody has ever thought."*
Albert Szent-Györgi

When Roberto Maglione first asked if I would write an Introduction for his developing book on the subject of Wilhelm Reich's atmospheric discoveries, I experienced several conflicting feelings, and many questions were raised. *Cosmic Orgone Engineering*, or *cloudbusting* as it is known in common language, can be one of the most awe-inspiring and potentially beneficial of Reich's many discoveries, and certainly it is one of the most controversial. It is additionally potentially dangerous for both the operator and the community in which it is undertaken, if not handled with considerable knowledge, care and skill. The responsibilities and demands for technical training in the cloudbusting method are similar in depth and rigor to what is required for the pilot of a large jetliner, or captain of a passenger ship. It requires years of study and apprenticeship, with critical evaluations and oversight, and is not for amateurs or the unserious. Issues are thereby automatically raised, as to the social benefits or detriments for such a book as this to appear, in a time of great social impulsiveness and rejection of all forms of authority, even when naturally-based upon extensive work experience and knowledge.

Most of those who possess exacting knowledge on this part of Reich's discoveries had for years spoken openly in public about the results of our work. The operational and technical details by contrast remained very private, as a subject for discussion only within groups of highly trusted associates, even while we remained quite open and very public on Reich's other controversial discoveries, such as the orgone energy accumulator and Reich's sex-economic findings. Mostly, this book follows a similar path, providing only the most tenuous of details on operational or technical matters. This is not a "do it yourself cloudbusting manual" which would by itself be socially irresponsible[1]; but it is an excellent introduction into a complex subject material.

Beyond this concern, I asked, will the book be true to Reich and to the difficult subject material? Will it retain an emphasis upon functional principles and the natural orgonotic self-regulatory process which demands that cloudbusting only be used as an occasional help for nature, and not as a chronic enterprise? Will it avoid an overly mechanistic or even mystical presentation of the subject, which for example is more typical of what is found on the global internet, littered as it is with the most awful distortions of Reich one can imagine. I am pleased to report that this new book has realized my hopes, and not my fears. Dr. Maglione has done an excellent scholar's job in summarizing the difficult details and positive accomplishments of this subject.

While I do not wish to oversimplify or exaggerate, from nearly 30 years of field work with the cloudbuster, I have come to appreciate the tremendous power and capacities of the underlying *orgone energy continuum* which the device stimulates. This same energy continuum governs our weather at its fundament, and also plays an important role in both biological and cosmological functions. It once was recognized as the *life-energy* by ordinary perceptive people within a hundred different cultures[2], and also used to be described in the context of the *cosmological ether of space* — which contrary to popular myth, was factually detected and proven repeatedly to exist by physicists, from the 1920s into the modern times.[3] And the cloudbusting method itself had been examined and proven scientifically years ago, in field studies reviewing official weather data and applying classical scientific methods of analysis. Dr. Maglione reviews this extensive evidence.

Personally speaking, I have seen cloudbusting operations which stimulated large rivers of atmospheric moisture to veer into the cores of hard drought regions, whereupon tired grey haze-clouds swelled up into large blue-white cumulonimbus, spreading rains across hardpan and cracked earth. I've seen regions dry for decades become significantly wet after only a few days of carefully-undertaken cloudbusting operations, with flowing creeks and streams, and reservoirs sometimes filled to overflowing. I've stood at the edge of some of the driest spots on the planet — the Kalahari, the Syrian and Negev Deserts, the Namib, and the great Sahara Desert — and in all cases, the cloudbusting method proved its worth.

Stagnant and hazy, overheated atmospheres were replaced, usually within a short stretch of a few days of hard work, with the most magnificent and mighty of rainclouds, pregnant with rain, and pouring forth life-giving moisture across the parched landscape. A dramatic blossoming of wildflowers, grasses and sometimes new lakes were observed in barren stretches of previously dead, dusty and parched land where only the most skimpy of rains had previously fallen, and where nobody could have believed it possible. In the parched "Holy" lands, suffering under the dual plagues of social and environmental disasters, members of our small group, now called the *CORE Network*[4], have witnessed dramatic transformations of the atmosphere from dry to wet, hot to cool, and with greened hills and lakes and snow-capped peaks appearing in places previously completely barren and arid. We had to pinch ourselves to be certain we were not dreaming, or had not been magically transported from the hard desert to some green and snowy-lush landscape typical of the Swiss Alps or Rocky Mountains.

During our most ambitious undertaking, a five-year project in Eritrea, Africa, large sections of the Sahel were greened and made verdant, and annual food imports to that small nation were reduced by tens of millions of dollars. Giant new lakes — too large to see the opposite shore — appeared as if by miracle within the heart of the Sahara, as overflow from the Nile, and after only a few summers of consistently-applied cloudbusting work. And yet, while ordinary people celebrated the results of our work, only in the most rare situation did we find support at the official level. Even with the largest and most beneficial results, with exceptional rains ending exceptional droughts, we were frequently subjected to attacks and hatred from the orthodox meteorologists and officials, as if we were practitioners of some kind of "black magic" they could not understand and wanted nothing to do with.

However, cloudbusting is definitely *not* "magic", but a combination of both natural science and empirical art, requiring the practitioner to know much about modern science, climate and technical matters. They must also have the capacity to *feel the atmospheric orgone energy via organ sensations*, and to see the expressive *language of the living* which appears across the whole of Nature, if one knows what to look for, and has the eyes to see. It helps us to understand previously inexplicable things such as the relationship between desert-spreading and the subsequent appearance of droughts and heat-waves, both of which fuel the misunderstood "global warming" and "El Niño" effects, which in fact are *regional* in nature, and always connected with outbreaks of expanding Saharasian desert atmospheres.[5] The new theoretical understandings which flow from Reich's discoveries give us more than mere insights into problems, however. They also provide new approaches to doing something constructive and positive about those problems.

As a consequence, we now have the technical means to stimulate a softening or ending of droughts and heat-waves, and to green the harshest deserts of the world, assuming that this is what humanity desires. Technically, I can say with confidence, this is now possible. We have this possibility within our hands. *Socially,* however, we still face tremendous and oftentimes insurmountable obstacles, given the emotionally-armored condition of modern *Homo normalis*, who loves and creates deserts far more rapidly than most anything else. It is hoped this book will create a strong wind to "clear the air" on a subject too-long submerged into smoke and confusion, and to clear away those obstacles. This is a book for the serious scholar, or for those with an interest to support serious work-efforts, or as a beginning text for the serious student who might ultimately wish to apply themselves towards the difficult work of greening deserts. This book by Roberto Maglione is timely, and fills an essential need in an era of continuing global desert expansion and severe drought.

James DeMeo, Ph.D.
Greensprings, Ashland, Oregon
September 2006

DeMeo Foreword Citations

1. DeMeo, J.: *"So You Want to Build a Cloudbuster?"*, Orgone Biophysical Research Lab, 1986
   http://www.orgonelab.org/sobuildaclb.htm -- also see: http://www.orgonelab.org/chemtrails.htm
2. DeMeo, J.: *The Orgone Accumulator Handbook: Construction Plans, Experimental Use and Protection Against Toxic Energy,* Natural Energy Works, Ashland, Oregon 1989.
3. DeMeo, J.: *A Dynamic and Substantive Cosmological Ether-Drift,* Proceedings of the Natural Philosophy Alliance, Cynthia Whitney, Editor, Vol.1, No.1, Spring 2004. Arlington, MA. p.15-20.
   Internet Posting: http://www.orgonelab.org/DynamicEther.pdf
4. CORE Network: http://www.cloudbusting.org
5. DeMeo, J.: *Desert Expansion and Drought: Environmental Crisis, Part I,* Journal of Orgonomy, Vol 23, No. 1, Orgonomic Publications Inc, New York, May 1989. DeMeo, J.: *"Desert-Drought Map"*, Pulse of the Planet #2, p.82, 1989. Also, Chapter on "The Saharasian Desert Belt" in DeMeo, J.: *Saharasia: The 4000 BCE Origins of Child-Abuse, Sex-Repression, Warfare and Social Violence, In the Deserts of the Old World,* Orgone Biophysical Research Lab/ Natural Energy Works, Ashland, Oregon 1998. See also the discussion and map on p.61-62 of this work.

Author's Introduction

*"Love, work and knowledge are the wellsprings of our life.
They should also govern it."*
Wilhelm Reich

In 1996 I was drawn to Wilhelm Reich by chance. I had never heard of him before. At the time, I was working in San Donato Milanese, Italy, and it was almost obligatory to go into the center of Milan after work. One day as I stopped outside a bookstore under the arcades of the Duomo square, I noticed a book in the window which spoke about Reich.[1] At that time, I knew absolutely nothing about him. The title made me so curious that I entered the shop and bought a copy.

I couldn't put the book down: it was about the research of an Austrian scientist, Reich in fact, in the field of natural sciences. During the American period of his life and work, he discovered an extremely simple method to make rain in arid and desert areas. The book also included an excellent bibliography.

I immediately went and bought everything that was available on his work and that had been written about him by other authors. I gained knowledge of psychology (as he was a psychiatrist, a student and great friend of Freud), biophysics, life energy, and of many other things he discovered during his long career. Gradually I read the literature, and as I involved myself in the science of orgonomy (which he had created), I began to realize that his approach was totally different from the one generally used: very simple, direct, immediate and without any frills. I had no difficulty in reading and understanding it. Particularly in understanding it. He was able to explain difficult and complex things in an almost obvious way, making them very simple.

I understood that Reich's research methods went much further than traditional scientific methods, but that in every case they always remained within the field of science, without ever leaving its boundaries. In addition, it was very interesting to note that he dealt with things that were generally difficult to approach with traditional scientific methods: the character structure of human beings; the significance of the orgasm, which he used to call the "Cinderella of Natural Science"; the origin of life and biogenesis; methods to cure illnesses based principally on the energy charge of the organism; the perpetual-motion machine and a free energy motor which he called *orgone motor*; changes in climatic conditions of an area using a particular system of tubes that he called *cloudbuster*; and many other things. It might all seem fantastic and surreal, the work of a dreamer and visionary. But it wasn't like that. In fact, how could a visionary carry forward his research, staying true to scientific methods? He published many books on the results he obtained, ranging from psychology and sociology to the study of the origin of life, and finally to astrophysics. He had close and trusted colleagues who shared his fundamental concepts over a period of nearly 40 years. Why was he so discredited and had to fight on so many fronts, practically against everyone, and was left to die in prison if he was only a visionary?

Something did not add up. Surely there had to be something more. I went deeper and discovered that I was not dealing with fantasies or conclusions picked from the air. On the contrary, these were the results of years of observations on firstly the behavior of human beings, and secondly, on natural phenomena. All in turn followed the same logical path: the existence of an energy that was always present in all the experiments he carried out and that is still unknown to traditional physics: life energy. An energy that, in past centuries, everyone spoke of and discussed, but no one was able to express in a concrete and quantifiable way.[2]

Reich was the first in the history of science who was able to measure and quantify the life energy, which he named *orgone*. He was able to identify and measure it in the soil, in the atmosphere, and in animal and plant organisms by means of a thermometer, which reveals the orgone by an increase in temperature; using the electroscope, as the concentration or density is expressed in the variation of the discharge of the electroscope; and the Geiger-Muller counter. On this matter, he wrote in one of his most important books, the *Function of the Orgasm*:[3]

> " ... *It was demonstrated that the sun emits an energy which influences rubber and cotton ... and the human organism after full respiration in the vegetatively undisturbed state. I called this energy, which is capable of charging organic matter, orgone....*
>
> *Orgone energy is also demonstrable visually, thermically, and electroscopically, in the soil, in the atmosphere, and in plant and animal organisms. The flickering of the sky, which some physicists ascribe to terrestrial magnetism, and the glimmering of stars on clear dry nights, are direct expressions of the movement of the atmospheric orgone...*
>
> *...The "electric storms" of the atmosphere which disturb electrical equipment during intensified sun-spot activity are, as can be experimentally demonstrated, an effect of the atmospheric orgone energy...*
>
> *.... The color of orgone energy is blue or blue-grey .... The spontaneous discharge of electroscopes in non-ionized air, a phenomenon designated as "natural leak" by physicists, is the effect of atmospheric orgone....*
>
> *.... The blue color of the sky and the blue-grey of atmospheric haze on hot summer days are direct reflections of the atmospheric orgone...*
>
> *.... The hitherto misunderstood formations of clouds and thunderstorms are dependent upon changes in the concentration of the atmospheric orgone. This can be simply demonstrated by measuring the speed of the electroscopic discharges. "*

I found out that much had been written on his work on character analysis, but very little on orgone physics, especially regarding some of the more ingenious and fascinating tools ever devised by man: the *orgone energy accumulator* and the *cloudbuster*. The *cloudbuster* in particular impresses with its simplicity and the variety of its application. It can be used both in the medical field to restore the body's energetic flow, as an instrument that is called the *DOR-buster*, and in meteorology to change and stimulate climatic conditions towards self-regulation, above all in areas that are subjected to drought or are in phases of desertification.

That is the reason that I decided to write something on this piece of equipment and on one of its most interesting applications: the restoring of rains to desert and dry meteorological and climatic conditions. I wanted to show what Reich did, but also what advances were made in subsequent years by other scientists, with a look at the modern applications as well. The result has been impressive. Not only did many people validate the theory Reich originally developed, but new protocols were developed and succeeded to bring increased rains to areas where it was never thought possible, like in Eritrea or the desert of Namibia, to name two examples. Also, all these experiments were developed in a rigorous scientific way, monitoring the atmospheric parameters and viewing the region by satellite, so as to create a picture as complete as possible, and to produce protocols and documentation that could be used for a replication of the experiments even by the traditional scientific community.

Therefore, the *cloudbuster* not only creates rain or dissolves clouds, depending upon how it is used, but can even influence heat, humidity and smog, and create climatic conditions more suitable and acceptable to life. It can also be employed in extinguishing fires that each summer destroy hundreds of hectares of forests and against which the traditional means have very little effect. It can even be used to divert the path of hurricanes. Kelley hypothesized that the use of the *cloudbuster*, as an instrument that can influence the meteorological conditions over a certain area, can even bring changes to a nation. In the preface of his report dedicated to *cloudbusting,* he states:[4]

> *"Weather provides the foundation for the existence of all life as we know it. Men and nations live and are shaped, or die, depending on the weather. A single rain can mean the difference between prosperity and bankruptcy to a farmer. A deluge can inundate an area the size of a European nation. A drought can disrupt a nation's economy, and bring hundreds of thousands to starvation. And a change in the climate can spell the end - or the beginning - of an entire civilization..."*

However, this instrument must be used with much caution, because if it is handled in an incorrect way, it could both cause irreparable physical damage to the operator, particularly to the cardiocirculatory system as well as to the body parts that are in direct contact with it, and to the region in which it is being used.

In fact, if one exceeds its use, it could produce unexpected atmospheric reactions that would not be easily reversed, possibly causing drought and desert-expansion, but also heavy hailstorms, thunderstorms, hurricanes, flooding, etc, posing a grave danger to everybody and everything in that area. It is therefore necessary that *cloudbusting* is performed by people who understand orgonomy and its laws very well and, most of all, who have had a great deal of experience in this field. I find it difficult to imagine that it will be possible to resolve great environmental problems such as drought, water shortage, and desertification simply by pointing a series of tubes at the horizon. It requires a great understanding of the laws that govern natural phenomena like the weather, and especially the basics of orgone physics. Otherwise there is a risk of creating even greater problems than those which our environment already faces.

All this is unfortunately basically possible since anybody, indeed everybody, can try to construct Reich's *cloudbuster* device, even only from hearsay, in so far as the laws and devices developed by Reich are actually fairly easy to build. This is a great advantage, but also has become a great problem in those cases where ignorant laypeople amateurishly use this knowledge purely out of curiosity, just to see if it works. In this case, there is a real risk of provoking abnormal atmospheric reactions and environmental disasters which put people and property in danger.

It is essential to act responsibly and leave the application of these devices to the authentic experts as well as to specialized research centers, to develop more secure and appropriate protocols for specific goals, so that *cloudbusting* can be used in a regulated and coordinated manner without such dangers.

Let's hope that in the near future, all this can be realized and the evolution of human beings and science can begin in a life-positive direction, as Reich had always wanted it to be, but whose realization in the short term had been very doubtful.

Let's hope that we will not have to read headlines like those that have appeared frequently in the press in the last few years and that are a testimony to increasing problems of an environment that is now in a state of miserable deterioration. Significant is the article that appeared in the Italian daily newspaper *La Stampa* on 10 May 2000, which read:

> *"Emergency agricultural plans ... No rain for eight months, grain and grapevine in danger ... Agriculture is on alert due to the present drought.*

"In Sardinia, where the planning of the water reserves has already started, all that they are able to do is to guarantee the survival of animals and plants. A line which will be difficult to maintain if the drought and the abnormal temperatures continue."

Or again, in *La Stampa* on 13 July 2002:

"... the advancing of the desert towards the north continues: it mainly concerns Spain, Sicily, Sardinia, Greece. Fires, heat, human activity and poor precipitation is changing the territory...

.... The water crisis of the south is dramatic, but the alarm has been extended to all of Italy and is serious even in the central parts. According to a survey by Coldiretti, a risky situation has emerged for industrial cultures such as tobacco. In Umbria – Coldiretti continues – Lake Trasimeno is below its critical threshold. Difficulties have been signalled in the Basso Tevere agricultural zone and the Orvietano, where crops suffering most are corn, tobacco, vegetables and beet ...."

And even more sadly significant are the news of the conditions that continue on the African continent and in Central Asia. The *International Herald Tribune* wrote the following on April 4th, 2002:

" ... Desertification has emerged as a significant danger not only for Spain, but also for the other three southernmost countries of the European Union, Portugal, Italy and Greece ... Desertification threatens all the other continents, as well. China and its capital, Beijing, are battling the worst dust storm in memory, and thick yellow clouds of fine sand swept into Japan and South Korea in March ...

.... More than a quarter of a billion people already are directly affected by desertification, according to the United Nations Secretariat of the Convention to Combat the Desertification in Bonn.

Another billion in more than 100 countries are at risk, including citizens of sub-Saharan Africa and Central Asia who are among the poorest people in the world ...."

We know that drought, water shortages, and an increase in the desertification make any form of life difficult if not impossible. In certain regions it has become a fight for survival, with a daily struggle against climatic conditions, hunger and illness. Measures of intervention for the restoration of normality in an area already hit, and prevention for those at risk are without a doubt desirable in the short term for the survival of this planet.

Chapter 1 Citations

1. Zabini, A.: *Wilhelm Reich e il Segreto dei Dischi Volanti*, Tre Editori, Rome, 1996.

2. In past centuries, countless scientists tried to explain the presence of a hypothetical life energy. The ancient Chinese already knew of the existence of an energy called *Chi,* which flows through the human body and the technique of acupuncture refers to. Several Indian books speak about a vital force called *Prana*. It runs through the body along the meridians, and drawings showed the major energetic points on the body of an elephant, in a manner very similar to Chinese acupuncture. Both cultures showed that this energy is absorbed by the organism through respiration, and flows inside it along the meridians. When this flow is blocked, the conditions are established to develop illnesses.

   In addition, the West also was developing similar philosophies advocating that an energy was present in living beings and in the cosmos. In 3 BC Hermes Trismegistus called it *Telesma*; Hippocrates, in the same period, gave it the name *Vis Medicatrix Naturae,* while Aristotle called it *Quintessence*. In 6 AD in Polynesia and Hawaii, a similar energy was called *Mana*. Later, in the 16th Century, Paracelsus called it *Munia*, and a century later, Kepler gave it the name *Facultas Formatrix*. Goethe started to speak in the 18th Century of a cosmic energy which he named *Gestaltung*, while in the same period, Galvani drew upon the concept of *Vital Energy*. A short time later, Mesmer talked of *Animal Magnetism* as an atmospheric fluid that surrounds, charges and enlivens all living things. A similar concept was also drawn up by the German Scientist von Reichenbach after numerous years of study and research. Seeing the presence of a natural energy which did not respond to traditional physical laws, he chose to call it *Odic Force*. Subsequently, Freud, studying the behavior of human beings, identified an energy that he was not able to completely quantify, naming it *Libido*, while a short time later, Jung called it *Synchronicity*. At the start of the last Century, the French philosopher and Noble prize winner Bergson drew up a theory of evolution based on the spiritual dimension of the human life. The basis of this process was a vital energy or *Elan Vital,* which was pure energy, free from physical or deterministic implications. In the same period, the Russian Scientist Lakhovsky identified the existence of a spiritual substance, infinitely subtle, intangible and impalpable, which penetrates all objects and living beings, extending into the universe and into interstellar space. He called this substance *Universion*. According to Lakhovsky, it is the basic substance from which all physical matter and living beings are derived by way of condensation.

   More recently Burr, from Yale University, came up with the presence of a potent natural electro-dynamical field that influences both meteorological conditions and human beings. The biologist Sheldrake developed a theory similar to that of Burr, that was based on an energy that he called a *Morphogenetic Field*. The French scientist Kervran, after numerous years of studying the transmutation of chemical elements in human beings, discovered that animals that have a diet high in silicate, transform that silicate into calcium. The results he got, confirmed by independent research in both Europe and Japan, suggest the presence of some unknown form of powerful biological energy which guides the transmutation. In the 1950s Abbot used theories based on concepts of energy flows in the atmosphere for weather forecasts. In this way, he was able to predict the weather with a high level of accuracy for months in advance. Among the physicists of the last century, Miller stands out, who proved the existence of a cosmic ether through his experiments. He also showed that the ether is dynamic and is reflected by metals, much like Reich had established. The Italian Chemist Piccardi studied the chemical physical behavior of water and showed that its properties can be influenced by a form of unknown cosmic energy, very similar to a strong magnet and correlated to sunspots. At the same time, the Russian Physicist Grishenko identified a stream of cool plasma as responsible for the energetic structure of the physical body of an organism, that he decided to call *Bioplasma*. Recently Inyushin, a researcher at the University of Kazakhstan at Alma Ata, Siberia, came to the conclusion that the bioplasma body is the same as the etheric body of the oriental philosophers and doctrines. The list of scientists who have studied and tried to clarify the natural presence of this life energy throughout the centuries, does not end here and could possibly be infinite, spanning all epochs. Aside from studying of the same energy, all these researchers also have in common that they have never been taken seriously by the academics of their time, even if, as in many cases, these scientists represented the top of the scientific establishment. In the end, they were almost always scientifically isolated and ridiculed, and the most unfortunate were assassinated or burned at the stake.

3. Reich, W.: *Die Entdeckung des Orgons, Erster Teil: Die Funktion des Orgasmus*, Internationaler Psychoanalytischer Verlag, Vienna, 1927 (English translation, *The Function of the Orgasm*, Farrar, Straus & Giroux, New York, 1961).

4. Kelley, C.R.: *A New Method of Weather Control*, Kelley/Radix Publications, Vancouver (USA), 1961. Also see Kelley, C.R.: *Eine Neue Methode der Wetterkontrolle*, Plejaden Verlag, Berlin (Germany), 1985. Also reprinted in Kelley, C.R.: *Life Force: The Creative Process in Man and in Nature*, Trafford Publishers, Victoria, BC (Canada), 2004.

Wilhelm Reich. Life and Work

"Friends, we are here to say farewell, a last farewell, to Wilhelm Reich. Let us pause for a moment to appreciate the privilege, the incredible privilege, of having known him. Once in a thousand years, nay once in two thousand years, such a man comes upon this earth to change the destiny of the human race. As with all great men, distortion, falsehood and persecution followed him. He met them all; until organized conspiracy sent him to prison and there killed him. We have witnessed it all, "The Murder of Christ". What poor words can I say that can either add to or clarify what he has done? His work is finished. He has earned his peace and has left a vast heritage for the people of this earth. We do not mourn him, but for ourselves, at our great loss. Let us take up the responsibility of his work and follow the path he cleared for us. So be it."

Elsworth Baker, Eulogy at Reich's funeral

Wilhelm Reich's life, and especially his work and research, can be divided into two distinct periods: the European period and the American period. In the former, which finished in August of 1939, Reich was mostly occupied with character analysis and biogenesis. Among the most important discoveries he made during this period are the sex-economic theory of human behavior, vegetotherapy, bions, and everything else related to the origins of life.

With the rise of Nazism, he was forced to move to the United States, where he remained until his death. In the American period, he developed orgone physics and laid the groundwork for the development and use of the orgone accumulator and the *cloudbuster*. In the spring of 1957, he was imprisoned in the Lewisburg penitentiary, convicted on a minor Food and Drug Administration (FDA) technicality. A large number of his books and instruments were burned and destroyed by FDA agents and court orders. He died in prison in November of 1957.

Numerous publications have been printed since his death, especially during the last thirty years, about his life and his scientific activities. Some of these were written by his family, like the biography by his second wife Ilse Ollendorff[1] which talks above all about his American period, or the book written by his son Peter[2], which describes his relationship with his father and his father's scientific activities in the last ten years of his life. Other were short booklets and interviews released by his daughter Eva[3] where she, aside from talking about her father and *cloudbusting*, extended his theories, developing new bioenergetic massage techniques and methods for childbirth.

In addition, students and close collaborators from both the European and American periods have documented their experiences from their times of common activities, and even scientists and researchers fascinated by the work and discoveries of Reich wanted to discuss and pass on his theories. I found such interesting texts like the one by Ola Raknes[4], who speaks of the Norwegian period and orgonomy; or by Edward Mann[5], who describes the orgone theory and its applications, demonstrating how it can be correlated to theories developed by other cultures during different historical periods, especially the Hindu-Yoga concept of Prana and the acupuncture theory of the Chinese.

Excellent biographies can be found, such as the one by Myron Sharaf[6], psychiatrist, student and collaborator of Reich, who talks about all aspects of the life and works of his teacher; by Jerome Greenfield[7] who deals mainly with the last years of Reich's life, with the lawsuit against Reich and orgone energy filed by the FDA; by Luigi de Marchi, psychologist, who first introduced a biography[8,9] and essays[10,11] on the works of Reich to Italy; up to the more recent work of Jim Martin[12], in which he discusses to what extent Reich's life was influenced by his slanderers and in particular by the international psychoanalytical organization, Nazism, Communism, the FBI, and the FDA[13]. All this together gave a complete view of Reich, and also helped to bring light into and to uncover some more information, both on his original research methods and on the curtain of conspiracy and silence that surrounded him and contributed to his death in prison.

The title Sharaf gave his book, *Fury on Earth*, is symbolic and indicates what really happened with Reich's life and his research: an authentic fury on the earth. He lived sixty years, he had three wives and lived with a fourth woman for many years. He had three children by two different mothers. He published more than 100 articles and 20 books, translated into many languages, and left several thousand unedited pages of manuscript, deposited today at the *Wilhelm Reich Museum* and legally sealed by his will for a period of 50 years (until 2007), which he himself stipulated as a date for making them publicly available. More than anything, he tried to shine a light on one of the most difficult and dangerous issues in the history of natural sciences, where many have started but almost all, for one reason or another, have failed: that of investigating the life energy. It seems a paradox: everyone who tries to explain how to use the energy which gives life and well-being, have been derided, destroyed, assassinated or sent to burn at the stake, and Reich was no exception. But he was the first who tried to identify and explain its presence and behavior within the cosmos through specific scientific methods, objectifying all his results and making the experiments repeatable wherever possible.

The European Period (1897-1939)

Wilhelm Reich, the first of two brothers, was born on the 24th of March, 1897 in Dobrzcynica, Galicia (which then belonged to the Austrian-Hungarian Empire and is currently part of the Ukraine), to secular Jewish parents. His family was well off and owned a farm of about 1,000 hectares. He spent his first years on the farm, where he was taught by tutors. As a young boy, he was showing distinct interest and curiosity for biology and natural sciences, in particular for the life processes of both plants and animals.

His mother committed suicide when he was about 11 years old. This had a profound influence on the rest of his life, both on his thoughts and on his work. His father died a few years later, when Reich was 17, and it fell to him to manage the estate, if only for a short period of time, as the Russians destroyed the farm and confiscated the land in 1915. Ending his studies in the same year, he joined the Austrian-Hungarian army as a second lieutenant. He fought on the Italian front until the end of World War I.

On his return from the war in 1918, he joined the faculty of Medicine at the University of Vienna, supporting himself in his studies mainly thanks to money he earned by giving private lessons. Soon, he became interested in psychoanalysis and began studying the works of Freud in depth. Within a short time, he became a member of the Viennese Psychoanalytical Society, two years before gaining his degree in Medicine. He then specialized in neuropsychiatry.

Reich's competence, both as an analyst and as an author of articles and essays on psychoanalysis, was appreciated by Freud to such extent that he offered him the position of First Assistant at Vienna's General Psychoanalytical Hospital. During that time, Reich married Annie Pink, a medical student, and they had two children, Eva (born 1924) and Lore (born 1928). In 1924, he began teaching at the Psychoanalytical Institute. He organized seminars, both at the institute and in the clinic where he worked as an assistant, focusing his courses and his research on what he considered to be the weaknesses in the approaches and practices of psychoanalysis at that time. In his analytical practice, the fact that all the patients under his care improved when they achieved a satisfactory sex life, made him very curious. This led him to conclude that, in order to cure the patients in an effective way, it would be necessary to overcome the stasis of the libido (which in that period was still considered an abstract concept) and for the patient to better regulate the internal flow of their organism's bioenergy.

He realized very quickly that mechanical sexual activity alone could not guarantee this. That it was, in fact, the gratification from the sexual act which would guarantee it.

Reich called this capacity for gratification *orgastic potency*. He arrived at the conclusion that the destructive behavior of human beings was not intrinsic, but came from the inability of obtaining sexual satisfaction. Until this time, psychoanalysts had considered sexual problems only as a symptom and not as the core and starting point of neurosis. They believed that the erective power of the penis was evidence of an adequate male sexual functioning. Recognizing the existence of *orgastic potency* was a great leap forward for the psychoanalytical movement at that time, and was a fundamental discovery within biology, as well as for the development of subsequent Reichian theories.

Essentially, *orgastic potency* signifies the capacity of the organism to discharge excess energy in order to maintain a stable energetic level within the organism. Reich noted that the metabolic process of the bioenergy followed a four-beat rhythm, composed of tension, charge, discharge and relaxation. He called this rhythm, which he considered to be fundamental in nature, the *orgasm formula*. In this way, Freud's concept of libido was transformed into something more than simply psychic or psychological. According to Reich, at the very basis of life was energy in its pure state.

Therefore, the neurosis exists only in cases of excess energy either being repressed or in stasis; a person who achieves a genuine capacity for sexual discharge will not develop neurosis. In addition, the person who is genitally healthy shows certain basic psychological characteristics, with a more relaxed and tolerant attitude towards society. At this point, ethics and many moral rules, drawn up for a civil life in society, become incomprehensible in the light of this new behavior. For example, living with a person you don't love, just because the law says that you are married, and guaranteeing a rigorous and coercive moral fidelity, makes no sense.

Nevertheless, the concept of morality did not disappear altogether for Reich. The rules concerning values are situated on a different level. For example, you desire a sexual relationship only when love is reciprocated; you lose interest in promiscuity; you consider pornography to be disgusting, and you are more tolerant towards others and society in general, with a more relaxed attitude. This is a self-regulated morality, as opposed to a compulsive one.

At this point the question arises, as it did for Reich: What happens when a person is taught, even from infancy, that sex is a prohibited thing, and consequently, the regular energy flow is stopped or blocked? What happens is that the pelvis is withdrawn, the muscles of the thighs and buttocks become rigid, the breath is held and the teeth are clenched, and one's attention is averted from everything that could restore the organism's natural energy flow. In the end, the sensation of sexual desire is lost and the body muscles harden like the strings of a musical instrument. Reich defines this condition as *being armored*. This process can continue until it involves all the muscles of the body, continuously increasing the energy level within the body. In the end, the excess energy transforms into neurotic symptoms. Generally, this process starts at birth, caused by the antisexual attitude of society. Very few young people can grow and develop according to the natural laws.

Substantially, for Reich, the orgasm regulates the body's energy flow while the muscular armor prevents the regulation of that energy.

Reich continually asked himself why and where that repression came from that strongly conditions human beings to suffer silently? Why was it necessary? And why was it so universal? These questions accompanied him throughout his life and deeply affected the direction of his research. He convinced himself more and more that the social causes were at the basis of the neurotic individual. In 1927 he published *Die Entdeckung des Orgons Erster Teil: Die Funktion des Orgasmus*[14], his first important book which detailed all of his discoveries, and in a particular explicitly discussed orgastic potency.

In 1928, he became vice-director of the psychoanalytic clinic in which he worked. He organized mental hygiene courses and established sexual guidance centers for young people, for workers and employees. It was the first time that psychiatric consultation and contraceptive information, children's education and sexual education was openly provided to adolescents and young people, to the working and middle class of society. He recognized the necessity for change in the social structure of society, and therefore he joined liberal and socialist groups. Reich believed they were genuinely open to social reforms. Freud did not approve of Reich's social efforts, however, where he continually mixed politics with psychoanalysis, and therefore the friendship between them began to slowly fade away.

In 1930, Reich moved to Berlin and, observing the rise of Hitler's power, joined the Communist Party, leaving his family in Vienna. In the same period he started a gradual separation from his wife, which increased to the point that they divorced a short time later.

In Berlin he organized and assumed the management of mental hygiene and sexual guidance centers. He pushed his ideas on social reform, which included better houses for the working class, the abolition of laws against abortion and homosexuality, and relaxing of strict laws regarding marriage and divorce. He worked to support health care for women and babies, and the setting up of nurseries in the work place and the poor districts.

In 1931, he published *Der Einbruch der Sexuellen Zwangsmoral*[15], where he tried to explain the origin of society's hostile attitude towards natural sexuality, based on Malinowski's[16] discoveries. All the clinical and ethnological material related to this text was collected between 1920 and 1930. In this book, Reich takes the first step towards addressing the neurotic problem of the masses, moving away from the ideas of Freud, who considered sexual repression to be one essential component in the development of society. Unfortunately, his ways of thinking and the way in which Reich brought forward his didactic activities were not in agreement with the Communist Party line. In fact, following his visit to the USSR in 1929, he became one of the most ruthless opponents of communist ideology, publicly identifying them as *red fascists*. In 1933, he was expelled from the Communist Party.

During this period he met the dancer Elsa Lindenberg, who was involved in underground anti-Nazi activities, and he lived with her until 1939.

In 1933, he published the first edition of *Charakteranalyse*[17], a classic on the understanding of the human character, in which he reviewed all the psychiatric work carried out since the 1920s. In this book, he also discussed the structure of the armored character and the laws that regulate the organism's energy flows. With this text, Reich opened a new era in psychoanalysis and in the approach to the biological basis to illness and neurosis, definitively uniting mind and body, psyche and soma. In the same year, he published *Die Massenpsychologie des Faschismus*[18], in which he discussed the genesis of the character structure behind the fascist ideologies and the connection between the personality of the individual and the fascist ideology. He showed how every form of organized mysticism, including fascism, is based on the sex-frustrations and orgastic longings of the masses.

In 1933, after Hitler had come to power, Reich left Germany and moved to Denmark, where he stayed for a short period of time. Because of pressure from the Nazis, he was then forced to move to

Sweden. There he was refused permission to stay, and following an invitation from the Psychological Institute of the University of Oslo, he moved to Norway in 1934.

In 1934, during the XIII International Congress of Psychoanalysis, held in Lucerne, Switzerland, he was expelled from the International Society of Psychoanalysis.

In 1936, he published *Die Sexualität im Kulturkampf: Zur Sozialistischen Umstrukturierung des Menschen.*[19] In this book he summarized the sexual conditions and conflicts of that time, showing how the institutions of matrimony and family can be considered the basis of spreading sexual repression.

During all those years, he tirelessly continued to carry forward his research, and to develop his theories and therapies on emotional problems. In Oslo, he started to study the bioelectric nature of pleasure and anxiety, searching to give an answer to the age-old enigma of the origin of life. Reich believed that when a person experienced sensations of pleasure, corresponding electric currents would form on the surface of the skin. Using a galvanometer, he started to investigate this and found experimental confirmation for his hypothesis. The more pleasure a person experienced, the greater the amount of current which could be measured with the galvanometer. On the other hand, in unpleasurable situations, the electric current disappeared. He concluded that during the sensations of pleasure, there is an expansion of the organism, evidenced with an increase in the peripheral electric current, while during anxiety the organism contracts, and the surface electrical current disappears.

Thanks to this experiment, he had concrete evidence for a real and measurable energy within the body, and he called it *bioelectric energy.* Later on he demonstrated that this energy extends beyond the surface of the skin in the form of an *energy field.* In this context, the genitals can be considered a special superficial organ capable of regulating the energy flow by discharging it from the organism. His term *sex-economy* describes the metabolic process of charging and discharging of this bioenergy. With these discoveries, Reich passed slowly from the field of psychology and psychoanalysis into biophysics, where he developed a totally new concept of health.

During his research to better understand bioelectricity, he studied the decomposition of food under the microscope, and found it broke down into tiny luminous vesicles that moved freely, and which could be cultivated. When germs or the cancerous cells were brought into contact with these vesicles, they were immobilized and killed. These vesicles appeared to be at an intermediary stage between living and non-living matter, and he called them *bions.* However, Reich wasn't the first to discover bions, nor to study them. Bastian[20], a contemporary of Pasteur, talked about them in his book *The Beginning of Life*[21], postulating that they were the origins of germs and illnesses. Yet the scientific community of that time accepted Pasteur's theory, which stated that germs couldn't change and always remain the same. Reich also observed that bions emitted radiation irritating to the eyes when viewed through a microscope, and that metal instruments positioned near the bion cultures spontaneously acquired a magnetic field.

He also noticed that a pair of rubber gloves left exposed to the sun, or to the bion cultures became highly charged. Moving them closer to the leaves of the electroscope, he saw that they deflected it in an exceptional way. He tried to shield this radiation by building a metal box that could reflect the radiation. Much to his surprise, he saw that in this box, the effects of the radiation greatly increased, and the effects could be noted even outside the box. It seemed that there was no way or possibility to protect oneself against this unknown radiation which appeared to come from everywhere. He demonstrated this energy came from the sun, from the bion cultures, and from living creatures, and later that it existed freely in the atmosphere. He found it everywhere, even in decomposing blood, tissue and grass. He confirmed that it was the same energy that he had found during the experiments on skin surface tension, and he decided to call it *vital energy,* and later, *orgone energy*, derived from the words "organism" and "orgasm". He concluded that it had a cosmic origin, a primordial energy from which all matter, animate and inanimate, originates. It was a great discovery which Reich considered second only to his prior discovery of the orgasm formula, of biological tension and discharge.

In the following years, he devoted himself to investigating and documenting this new type of bioenergy, including its physical and biological properties.

He demonstrated that cancerous cells developed from the decomposition of living tissues under energetically stagnated conditions, and that cancer is a biopathic consequence of sexual repression, which brings about emotional resignation and bioenergetic death of the organism.

In 1938, he published *Die Bione*[22] in which he reported the results of the experiments carried out in Oslo in 1936/37, and where he applied the orgasm formula to microbiology.

The American Period (1939-1957)

In 1939, an ever-increasing number of psychiatrists from the University of Oslo disagreed with Reich's work, research methods and conclusions, and proceeded to impede and publicly denigrate his work. The situation became unbearable to the point that Reich accepted the invitation from Theodore Wolf, his student and good friend, to give lessons on medical psychology at the *New School of Social Research* in New York. In August of the same year, he moved to the United States and set up a practice in Forest Hills. On the 20th of December 1939, he married Ilse Ollendorff who became a trusted assistant in his work, and remained so even after their divorce in 1951. They had a son, Peter, who was born in 1944.

Consistently following his discoveries, he came to the realization that the armored character structure was formed due to specific blocks of bioenergy flow inside the organism. He discovered that the character structure of disturbed therapy patients would spontaneously change towards non-neurotic conditions without the use of sophisticated psychological techniques, by directly freeing and enabling the natural flow of the biological energy blocked in the organism.

He identified seven segments into which the organism could be divided, each one mostly independent from the others, but at the same time, interdependent as far as the unified function of the organism is concerned. These segments — the ocular, oral, cervical, thoracic, diaphragmatic, abdominal, and pelvic — were of fundamental importance in the body's natural energy flow and functioning, but also participated in the development of the muscular armor under disturbed bioenergy conditions.

In those years, he proceeded to educate young doctor on his cancer research and orgone energy. In particular, he undertook many experiments with the orgone energy accumulator on animals and human beings. The accumulator, a cubic box very similar to a Faraday cage and capable of concentrating orgone energy on the inside, was built from alternate layers of organic and metallic materials. Reich observed, on the inside there was a temperature difference of +0.3 to +0.4°C compared to the temperature outside.

He carried out experiments with the accumulator with both healthy and cancer mice. He saw that the mice suffering from cancer lived longer when treated in the accumulator than the untreated cancerous control group. He noticed that when aluminium was used as the metallic material for the construction of the accumulator, it caused the mice to lose their fur when they were treated. The best results were obtained when he used iron sheets. Experiments on human beings showed in most cases an improvement of the general state of health, which Reich attributed to an increase in the organism's otherwise low energy level. In organisms which already contained high levels of energy, the use of the accumulator was not advisable, however, due to the problem of overcharge. He deduced that even illnesses such as depression could be symptomatically helped with the accumulator, as the energy level of such a person is low.

The results of the experiments were first published in the *International Journal of Sex-economy and Orgone Research*, published from 1941 to 1945, and later in the *Orgone Energy Bulletin* that was published from 1949 to 1953.

In early 1941, he showed the results obtained with the accumulator to Albert Einstein, who taught at Princeton University, New Jersey, at the time. At first, Einstein confirmed one of Reich's results, on the spontaneous temperature-difference inside the accumulator, to be a real and truly revolutionary discovery in the field of physics, since it experimentally went against the laws of thermodynamics. Later, Einstein changed his mind and rejected Reich's discovery without submitting the accumulator to more important controlled experiments. The motive for his behavior was probably fear of seeing his scientific work as a physicist compromised, in so far as Reich's theories upset traditional physics to the point that even Einstein's own work would be affected. Reich published the correspondence between them in a booklet with the title *The Einstein Affair.* [23]

In 1946, he gained American citizenship. The following year a new defamatory campaign was launched, based on distorted magazine articles and rumors about the sexual aspect of his research. The FDA then started an investigation into his research activities. After this, even the *American Psychiatric Association* and the *New York Psychoanalytic Association* sided against his work, but without substantive review.

Reich continued his research on the potential applications of orgone energy, discovering that it could be used even to power a motor. He devised a motor that functioned on orgone energy. The energy was obtained through the excitation of an orgone energy accumulator with a potential of 0.5 volt, needed to start a motor of 25 volts. Since the energy could not be stored, the motor had difficulty functioning in humid weather, when the absorption of energy from the environment is greatly reduced. Unfortunately, the research into this field didn't continue, and even one of the few prototypes that existed was stolen by one of his assistants who disappeared without trace. Reich noticed that this type of motor rotated much faster and with much less noise than the traditional motor, and he speculated that it could be used as a propulsion method for space ships, for travelling between planets. In 1948, he published *The Cancer Biopathy* [24] in which he reported very concisely on the results of his laboratory research from 1938 until the mid 1940s. In this text, he highlighted both the applied research methods that led to the discovery of the orgone energy, and the approaches which followed to explain the etiology, treatment and prevention of cancer.

In 1948 Reich was able to produce the lumination of orgone energy, concentrated in a high vacuum glass tube. He demonstrated that the orgone energy, which also existed in a vacuum, could also exist in space. In the same year, he formed the *American Association for Medical Orgonomy* and held the first International Congress of Orgonomy at Orgonon, Maine, his research center established in 1942. During this period, he also published *Listen, Little Man* [25], a vivid critique on the behavior of the average man.

In 1949, he founded the *Orgone Institute* and established the *Wilhelm Reich Foundation*. With these organizations, he planned additional research on cosmic orgone energy and its applications in the scientific field.

In 1951, he published *Ether, God and Devil* [26], where he spoke of *orgonomic functionalism* and *Cosmic Superimposition* [27], and how man is deeply rooted in nature. He spoke for the first time of the superimposition of two energy streams as one common functioning principle which was repeated infinitely in nature, and was the origin of matter. In fact, he established that the spiralling movement of two or more streams of mass-free orgone energy attract each other and merge, thereby creating a particle of inert mass. He called this phenomenon *bioenergetic function of two orgonotic systems*. He also recognized that the concept of superimposition was the basis for all natural phenomena like hurricanes, the formation of galaxies, the Northern Lights, and gravity. From this period also comes the monograph entitled *The Orgone Energy Accumulator* [28], a booklet in which he described the basic theory of the function of the orgone energy accumulator and its uses in the medical field.

At the beginning of 1951, he started research into the effects of orgone energy on nuclear radiation.

He undertook the *Oranur experiment*, a term derived from *Orgone anti Nuclear* (Or-a-Nur), based on the hypothesis that orgone energy could overcome the effects of nuclear radiation sickness. He obtained an opposite result: nuclear energy showed to be very noxious to orgone energy.

On the 5th of January 1951, he put 1 mg of radium, kept in its original lead container, into an accumulator with 20 layers, and he left it inside for five hours. This protocol was repeated daily for a week. On the last day, he left it inside the accumulator for only half an hour with dramatic consequences. The needle of the Geiger-Muller counter rose in an alarming way and often remained stuck at high levels. The buildings and the surrounding atmosphere showed a dim blue glow, and a characteristic odor pervaded in the atmosphere. Dull grey clouds formed over all the area. One doctor almost lost his life, while Reich himself fell ill and remained between life and death for some days.

The results of this experiment were described in the book *The Oranur Experiment, First Report (1947-1951)*[29]. Despite the danger and the experiment's disastrous results, Reich established that orgone energy could actually be turned into something deadly, which he called *Deadly Orgone* or DOR. It was the result of the nuclear radiation's effect on the orgone energy. DOR is essentially exhausted and immobilized orgone energy, which has lost all its vital charge during the contact with nuclear radiation. It is dark, almost black, toxic, and without radiance, and hungry for oxygen and water. It appears in the atmosphere like a grey haze or blackish flecks. The sun can be clearly seen through it, but photographs appear as if they had been taken in the shade. In the presence of DOR, tree leaves tend to loose their brightness, birds and insects stop and fall silent, and even air becomes stagnant. Human beings feel uncomfortable, the mouth dries out and they feel thirsty, and headache and nausea can develop. Reich called this phenomenon *DOR-sickness*. An organism that comes into contact with DOR for a certain period of time develops this illness. This is different to the *Oranur-sickness*, that Reich defined instead as a state of constant hyper-excitation of the organism, caused by the continued reaction of nuclear energy with orgone energy, and that precedes the formation of *DOR-sickness*.

It was in this period that, to eliminate the DOR that dominated the area of Orgonon, he invented the instrument that emphasized Reich's genius more than anything else: the *cloudbuster*. Elsworth Baker[30], a close co-worker, told in the *Journal of Orgonomy*[31] of the experience that caused Reich to build the *cloudbuster*. He wrote:

> *".... Early in 1952, to combat DOR, Reich devised a cloudbuster consisting principally of hollow pipes grounded in water, which was able to draw this energy out of the atmosphere into water. This led to an interest in weather, and he experimented with both the production and prevention of rain. On two occasions, his weather work was televised, and once, in 1953, he produced rain for a group of Maine farmers whose crops were dying because of drought. They had agreed to pay him if he produced rain within a prescribed time, and did so when he achieved this. In the early days of cloudbusting, Reich used five pipes, two above the other three. On one occasion, I saw him point the cloudbuster at a heavy cloud, and in a few moments, five holes appeared in this cloud, two above, and three below. I was convinced ..."*

In the same period, he developed his theory of desertification and drew up a project to restore the atmospheric self-regulation for charge (clear) and discharge (rain), creating favorable conditions for the natural generation of rain and the growth of vegetation in desert areas, or regions greatly at risk from desertification.

He also saw that the same principle, applied for the restoration of the pulsatory atmospheric cycle, could be employed for the treatment of neuroses. He modified the *cloudbuster* for use upon human beings, devising the instrument known as the *Medical DOR-buster*. The energy blocked in some parts of the organism due to muscular tension, could be re-vitalized with the use of the *DOR-buster*. He saw that, as soon as the energy-blocks had been diminished, the individual once again showed their natural emotions and functions.

Reich's discovery of orgasm anxiety and the armored character explains many enigmas concerning the behavior and thoughts of human beings. One of the most important was called the *emotional plague*. This type of highly armored and disturbed human behavior has the tendency to block all progress tending towards a natural functioning of life. Reich stated that no one is completely free from this illness and everybody is affected to some extent. Yet there are certain people who behave, from an emotional point of view, totally like the plagued. Generally these individuals are very intelligent and energetic, but with a strong push towards antisexual attitudes, and towards acquiring positions of authority in a way so as to dictate the rules of life, morals and ethics to everyone else.

In 1953, Reich published *The Emotional Plague of the Mankind*[32], composed of two volumes: *The Murder of Christ* and *People in Trouble*. In the first volume he spoke of Jesus Christ as a young and strong person, who did not absolutely profess chastity and asceticism, describing him instead as the symbol of sensuality and genital character. According to Reich, this was the reason for the murder of Christ. In the second volume, he reflected on his own participation in the Marxist movement around 1930. He explained how this experience led him to understand how character structure influences social processes.

In the years from 1945 to 1953, he made notable discoveries in the field of orgonomic medicine, orgone physics and mathematics. He developed a totally new method for the study of cosmic functions and gravity, basing it primarily on the results of the *Oranur* experiments.

However, he was not left in peace to do his work as he was constantly smeared and attacked both by the medical community and in newspapers and magazines. All this culminated in a formal Complaint from the FDA which he received on 10 February 1954. The District Court of Portland, Maine, ordered him to appear in court on the basis that the orgone energy accumulator was "mislabeled merchandise", that orgone energy didn't exist, and that all his books and journals on orgone energy research were "advertising literature" meant to promote the sales of the orgone accumulator. Reich did not appear in court, but sent a written Response to the judge, stating that the law could not judge on issues of natural science and scientific research, and that the work and research done by him fell into that category. The judge ignored Reich's written Response, and issued a court injunction for the 19th of March. The FDA, without any proof of facts and without having ever carried out any serious experiments to evaluate the orgone accumulator, was thereby successful in getting the District Court to ban the accumulator, asserting that orgone energy didn't exist and that all the supporting literature had to be burned. It became illegal for Reich to either sell accumulators or his books.

Despite these facts, he continued his research and publishing. In October 1954, he moved to Tucson, Arizona, where he used two *cloudbusters* to verify if it was possible to bring unusual rains and overcome desertification in that area. He stayed there until April 1955. This experience was reported in his last book entitled *Second Oranur Experiment: Contact with Space,*[33] published in 1957. In this book he discussed the forming and the increasing of the desertification of our planet.

In that period, he also came close to the solution of negative gravity. He postulated that gravity was due to a reaction of two overlapping energy streams, following all the general rules of cosmic superimposition. In 1954, the journal *Orgone Energy Bulletin* ceased publication, and the *Cosmic Orgone Engineering* (*CORE*) journal was published in its place. Reich returned to Orgonon in May 1955, and in the same summer held his last orgonomy conference, centered on the medical use of the *DOR-buster*. During this seminar, he met Aurora Karrer who became his third wife.

On July 26th 1955, Reich and one of his co-workers, Michael Silvert, were charged with *Contempt of Court*. Silvert had moved a shipment of books across the state border from Maine to New York, in violation of the original FDA injunction against "Interstate transportation of mislabeled merchandise". The trial was held in May of the following year, and both men were found guilty. Reich was sentenced to two years in prison, Silvert to one year and a day. The *Wilhelm Reich Foundation* was given a fine of $10,000. In the end, the accumulators and publications stored in a New York warehouse, and at Orgonon in Maine, were destroyed. Reich said that he would not be able to survive prison because he suffered from cardiocirculatory problems. On the 12th of March 1957, he was imprisoned in the penitentiary of Danbury, Connecticut, and later transferred to the penitentiary of Lewisburg, Pennsylvania. He died in prison on 3 November 1957, a week before he was due to be released.

While he was in prison, he may have solved the problem of negative gravity, as written in his last book, *Creation*. However, the manuscript for *Creation* disappeared mysteriously from the prison, and the formula of negative gravity probably died with him. He was buried at Orgonon on 6 November 1957. As per his will, and in spite of the financial costs of the trial, the lights in his study at Orgonon remained continually lit for one year.

Also, his will contained the stipulation that his archives, containing many documents and much unedited material, should only be opened and made public 50 years after his death. Therefore, after November 2007 we will be able to access a vast number of documents that could explain many of the mysteries of nature which Reich had not wanted to make public before, and especially some things that he had not fully revealed during his lifetime, for instance the orgone motor and some aspects of biophysics, like orgonomic mathematics and possibly anti-gravity.

After his death, some volumes of unedited material regarding mostly some of his diaries and interviews were released. And thus, *Reich Speaks of Freud*[34] was published in 1967, in which he talks about Freud, his great teacher, and of the years when he moved away from his theories of traditional psychology and towards the development of orgonomy.

Also, in the last 10 years of the 20th century, three interesting volumes were published by the executor of his will, Mary Boyd Higgins, revealing Reich as a man apart from his role of scientist, and much about his private life. Autobiographical texts, sometimes already partially published during his lifetime, were combined with materials from his diaries, work notes, and published articles from his journals, to give us a more homogeneous and complete view both of the adolescent and young Reich[35], and of the European[36] and American[37] periods.

Chapter 2 Citations

1. Ollendorff, I.: *Wilhelm Reich. A Personal Biography*, St. Martin's Press, New York, 1969.
2. Reich, P.: *A Book of Dreams*, Pan Books Ltd, London, 1974.
3. Reich, E.: *Orgonomic First-Aid for Mother and Infants*, Flatland Books, Fort Bragg, California; *Part 1: Wilhelm Reich and the UFO Phenomena: Desert OROP Ea*, VHS Tape, Flatland Books, Fort Bragg, California, 1995; *Part 2: Gentle Bio-Energetics: Prevention of Neurosis from Birth on*, VHS Tape, Flatland Books, Fort Bragg, California; *Teaching Gentle Baby Massage*, VHS Tape, Flatland Books, Fort Bragg, California.
4. Raknes, O.: *Wilhelm Reich and Orgonomy*, St. Martin's Press, New York, 1970.
5. Mann, E.: *Orgone, Reich & Eros. Wilhelm Reich's Theory of Life Energy*, Simon & Schuster, New York, 1973.
6. Sharaf, M.: *Fury on Earth. A Biography of Wilhelm Reich*, Da Capo Press, New York, 1994.
7. Greenfield, J.: *Wilhelm Reich vs. the USA*, WW Morton & Company, New York, 1974.
8. De Marchi, L.: *Vita ed Opere di Wilhelm Reich. Volume I. Il Periodo Freudiano-Marxista (1919-1938)*, SugarCo Edizioni, Milan, 1981.
9. De Marchi, L.: *Vita ed Opere di Wilhelm Reich. Volume II. La Scoperta dell'Orgone (1938-1957)*, SugarCo Edizioni, Milan, 1981.
10. De Marchi, L.: *Wilhelm Reich. Biografia di un'idea*, SugarCo Editore, Milan, 1973.
11. De Marchi, L.: *Wilhelm Reich. La Teoria dell'Orgasmo ed Altri Scritti*, Lerici Editori, Milan, 1961.
12. Martin, J.: *Wilhelm Reich and the Cold War*, Flatland Books, Fort Bragg (USA), 2000.
13. Acronym for *Food and Drug Administration*, American governmental department for regulation of foods, drugs and cosmetics coming into the American Market.
14. Reich, W.: *Die Entdeckung des Orgons Erster Teil: Die Funktion des Orgasmus*, Internationaler Psychoanalytischer Verlag, Vienna, 1927 (English translation, *The Function of the Orgasm*, Farrar, Straus & Giroux, New York, 1961).
15. Reich, W.: *Der Einbruch der Sexuellen Zwangsmoral*, Verlag für Sexualpolitik, Berlin, 1931 (English translation, *The Invasion of Compulsory Sex-Morality,* Farrar, Straus & Giroux, New York, 1961).
16. Bronislaw Malinowski was born in Krakow in 1884 into an aristocratic family. He was interested in anthropology and founded the discipline of social anthropology. In 1908, he attained a Ph.D. in Philosophy, Physics and Mathematics at the University of Krakow, and in 1913 one in Science at the *London School of Economics*. In the years 1915 to 1918, he undertook an expedition to the Trobriand islands in New Guinea, in the southwestern Pacific, in order to study the habits and ways of life of the indigenous people of that area. He always used a holistic approach in his studies, observing that all components of a society interact to form a balanced system, and he noted that the well-known theory of the Oedipus complex, which Freud had claimed to be universal, did not hold true in that non-European cultural context. He demonstrated that in a society such as that of the Trobrianders, which was characterized by great sexual freedom within a system of parentage and family dynamics organized in a matrilineal rather than patrilineal manner, the Oedipus complex would not develop and in fact could not be found. This proved that the psychoanalysis of the individual, such as Freud understood it, principally depended on the cultural context. In 1924, he started to teach at the University of London where he obtained a chair in Ethnology in 1927. He then undertook an expedition to Africa to document and study African tribal habits. He visited Oaxaca Valley in Mexico in 1941/42. Thanks to his experiences, gained from different populations situated on different continents, he was able to perfectly integrate culture and psychology. He demonstrated that culture was not only a system of collective habits, but also possessed specific social-institutions which play a role in its functioning. In each type of civilization, objects, ideas, and beliefs fulfill a certain vital function and represent a part of a whole always at work. He died in 1942 in New Haven, Connecticut, where he had moved for a period of time to teach at Yale University. He wrote a number of books which included *The Trobriands Islands* (1915), *The Scientific Theory of Culture* (1922), *Argonauts of the Western Pacific* (1922), *Crime and Custom in Savage Society* (1926), *Sex and Repression in Savage Society* (1927), *The Sexual Life of Savages in North-Western Melanesia* (1929), and posthumously *The Dynamic of Culture Change* (1945) and *Magic, Science and Religion* (1948).
17. Reich, W.: *Charakteranalyse*, Sexpol Verlag, Copenhagen, 1933 (English translation, *Character Analysis*, Farrar, Straus & Giroux, New York, 1949).
18. Reich, W.: *Die Massenpsychologie des Faschismus*, Sexpol Verlag, Copenhagen, 1933 (English translation, *The Mass Psychology of Fascism*, Farrar, Straus & Giroux, New York, 1969).
19. Reich, W.: *Die Sexualität im Kulturkampf: Zur Sozialistischen Umstrukturierung des Menschen*, Sexpol Verlag, Copenhagen, 1936 (English translation, *The Sexual Revolution,* Orgone Institute Press, New York, 1945).
20. Charlton Bastian was born in Truro, Cornwall, on April 26th, 1837. As a neurologist and scientist, he was mainly concerned with biogenesis and the problems associated with medical pathologies. He graduated in Medicine from London University in 1856. The year after, he became a professor of Anatomical Pathologies at the University College of London, and from 1887 to 1898, he held the office of the Head of Medicine. He was a member of the *Royal Society* of London from 1868;

consultant at the *Royal College of Physicians* in London from 1897 to 1898; an honorary member of the *Royal College of Physicians of Ireland*; and received a honorary degree in medicine from the *Royal University of Ireland*. From 1884 to 1898, he acted as consultant to the English Crown for cases of presumed mental dementia. He maintained an intensive scientific activity and he was a very prolific writer. In 1868, he published a series of articles about language disturbances. He dedicated much of his time to abiogenesis. He believed, contrary to the opinion of the time on biology and bacteriology, that a direct link existed between organic life and the inorganic realm. He refuted the accepted thesis of the scientific community of the time of *"omne vivum ex nove"*, stating that living matter must come from non-living matter at earlier stages of evolution, in a continuously ongoing process. He also disproved the belief that boiling water would destroy all bacteria it contained, as claimed by Pasteur, opening a road into the study of heat-resistant spores. He studied the origin of life, believing in spontaneous generation, and that living things came from inorganic matter, or from dead animals, or from plant tissues, through new molecular combinations. His theories and the results of his experiments are collected in a book of over 1100 pages, titled *The Beginning of Life* (1872). However, his theories were never shared by the scientists of his time: Pasteur (1822-1895), Tyndal (1820-1893), and Koch (1843-1910). He died on November 17th, 1915 at Chesham Bois, in Buckinghamshire. Among the most important books written by him, we find *The Modes of Origin of Lowest Organism* (1871), *The Brain as an Organ of Mind* (1880), *A Treatise on Aphasia and Other Speech Defects* (1898), *The Nature and Origin of Living Matter* (1905), *The Evolution of Life* (1907), and *The Origin of Life* (1911).

21. Bastian, C.: *The Beginning of Life*, Appleton, New York, 1872.

22. Reich, W.: *Die Bione*, Sexpol Verlag, Oslo, 1938 (English translation, *The Bion Experiments*, Farrar, Straus & Giroux, New York, 1979).

23. Reich, W.: *History of the Discovery of the Life Energy (American Period 1939-1952). The Einstein Affair*, Orgone Institute Press, Rangeley, Maine, 1953.

24. Reich, W.: *The Cancer Biopathy. Volume I & II*, Orgone Institute Press, New York, 1948.

25. Reich, W.: *Listen Little Man*, Orgone Institute Press, New York, 1948.

26. Reich, W.: *Ether, God and Devil*, Orgone Institute Press, New York, 1951.

27. Reich, W.: *Cosmic Superimposition*, Orgone Institute Press, Rangeley, Maine,1951.

28. Reich, W.: *The Orgone Energy Accumulator. Its Scientific and Medical Use*, Orgone Institute Press, Rangeley, Maine,1951.

29. Reich, W.: *The Oranur Experiment, First Report (1947-1951)*, Orgone Institute Press, Rangeley, Maine, 1951.

30. Elsworth F. Baker, born in Summit (North Dakota) in 1903. Psychiatrist, student and co-worker of Reich. He was the president of *the American College of Orgonomy, the American Board of Medical Orgonomy, the Orgonomic Research Foundation* and *Orgonomic Publication*. He published *Man in the Trap* (MacMillan Publishing Company, New York, 1967) in which he discussed and developed Reich's ideas on character analysis and sex-economy, according to which human character is based on the movement and on the blocks of the sexual energy. He died in New York in 1985.

31. Baker, E.F.: *Wilhelm Reich*, Journal of Orgonomy, Vol. 1, No. 1&2, Orgonomic Publications Inc, New York, 1967.

32. Reich, W.: *The Emotional Plague of the Mankind. Vol. I – The Murder of Christ*, Orgone Institute Press, Rangeley, Maine, 1953, *Vol. II – People in Trouble*, Orgone Institute Press, Rangeley (USA), 1953.

33. Reich, W.: *Contact with Space*, Core Pilot Press, Rangeley, Maine, 1957.

34. Reich, W.: *Reich Speaks of Freud*, Farrar, Straus & Giroux, New York, 1967.

35. Reich, W.: *Leidenschaft der Jugend,* Wilhelm Reich Infant Trust, New York, 1988 (English translation, *Passion of Youth: An Autobiography 1897-1922*, Farrar, Straus & Giroux, New York, 1988).

36. Reich, W.: *Jenseits der Psychologie*, Wilhelm Reich Infant Trust, New York, 1994 (English translation, *Beyond Psychology: Letters and Journals, 1934-1939*, Farrar, Straus & Giroux, New York, 1994).

37. Reich, W.: *American Odyssey: Letters and Journals, 1940-1947*, Farrar, Straus & Giroux, New York, 1999.

Traditional Methods of Weather Modification

"Silence is the most powerful weapon of EVIL...."
Maurice Magre

Human activity has always been influenced by meteorological events. For this reason, since ancient times, man has watched the sky, observed animal behavior and trusted in various beliefs to determine the weather for the following days.

For the same reason, the changing of the existing climatic conditions to his own advantage has been a necessity that man has brought with him since ancient times. It was a need caused by the scarcity or total lack of water, necessary for the vital and agricultural functions. At the time, the concept of modifying the weather could basically be reduced to that of rain making. The methods used to produce rain were not really scientific and mostly relied on religious or shamanic practices. They were generally performed by people of a certain authority in the local communities in which they lived, such as witches, shamans and psychics.

With the beginning of science and its discoveries, and with the extended studies of the mechanisms that underlie atmospheric phenomena, it became a more systematic and detailed vision.

The first in recorded history to carry out regular observations of atmospheric events were the Chinese in the 13th century B.C. In the west, the Greeks were the first to be interested in meteorology and to study atmospheric processes. Aristotle dealt in his book *Meteorologia*[1] with practically all atmospheric and geological phenomena of nature: rain, wind, lightning, typhoons, floods, climatic changes, earthquakes, the milky way, comets, shooting stars, etc. Due to its scope, Aristotle's text had a great influence on the development of meteorological studies among his contemporaries, was in widespread use even in the following centuries, and served as a reference in the Western world for long afterward. Even in modern times, authors such as Cartesio or Bernardino Telesio have cited the principles and theories it contains.

During the Renaissance period, between the 17th and 18th centuries, a number of measuring instruments were developed, among them the mercury thermometer, the barometer, and the heliothermometer. The latter was used by Fourier[2] to measure atmospheric warming and the greenhouse effect. At the end of the 19th century, Arrhenius[3] had identified the atmospheric properties which gave rise to these measurable effects.

Bjerknes[4] is counted among the most important meteorological scientists of the last century. He was one of the first to do in-depth research into the flow of air masses in the atmosphere. After that, Bergeron[5] began his studies on the precipitation mechanisms, while Solberg[6], in collaboration with Bjerknes, explored the phenomenon of cyclones.

With the passing of time and the gradual development of scientific knowledge, ways to concretely forecast meteorological conditions started to be investigated, in order to develop forecast methods that were more trustworthy and valid for longer periods of time.

In more recent times, we have began to think about how to modify the climate to the advantage of a certain area. To this end, many studies for making rain, producing artificial snow, combating hail, diverting hurricanes, and generally creating or modifying a certain meteorological situation have been undertaken. However, weather modification is not easily accomplished, as the atmosphere is a very complex and chaotic, even living system, powered by a massive amount of energy coming from the sun in particular. Its behavior is governed by a group of very heterogeneous rules, which in turn depend on a multitude of physical parameters and variables like wind velocity, humidity, temperature and atmospheric pressure.

The earth's atmosphere has been divided into different zones, according to the vertical distribution of the temperature. Figure 1 is a schematic representation of the atmospheric zones around the globe that are closest to the ground. The numbers show the minimum and maximum distance according to the latitude, season and meteorological conditions.

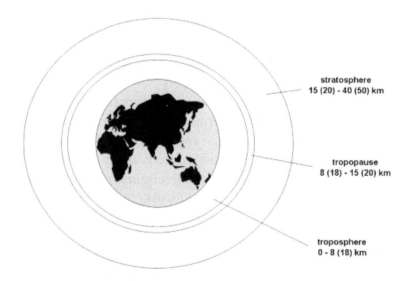

stratosphere
15 (20) - 40 (50) km

tropopause
8 (18) - 15 (20) km

troposphere
0 - 8 (18) km

Figure 1: Distribution of Atmospheric Zones

Despite the chaotic nature of the atmosphere, scientists today are inclined to think that climatic conditions cannot only be modified, but even that it is possible to have a global control of the climate. According to the experts, the solution to the problem is to be able to introduce small quantities of energy into certain areas precisely at the right moment, in a way to cause the desired effect. Naturally, all of this can only be done if the exact behavior of the atmosphere is known in its entirety.

Although it is behaving in a chaotic way, its behavior can be reliably simulated thanks to the scientific study of chaos, which is already successfully applied in other areas such as the orbit of space ships or the generation of radio signals. In addition, great progress in the understanding of the atmosphere could be achieved through the NASA satellite GIFT[7] which was put into orbit in 2004. It supplies detailed maps of the temperature trends and the concentration of water vapor present. Based on this information, it can therefore be decided to introduce into the atmosphere, at the right moment, the quantity of energy necessary to cause the planned event.

However, the experts warn that the meteorological conditions in a determined area can certainly bring benefits to that area, but can also produce undesired effects somewhere else. The circulation of air masses and climate are interdependent, and are parts of the same mechanism that determines and regulates the general atmospheric circulation. So if the climatic conditions are altered, at whatever point, there will certainly be an effect in another area that could also be unfavorable.

It is also important to emphasize that the weather modification on a large scale has both social and economical implications that must be considered, both locally and on a larger scale, before undertaking any such activities.

The statement made by Lorenz from the *Massachusetts Institute of Technology,* which was mostly interpreted as a joke, is therefore essentially true. He said that the flapping of the wings of a butterfly in Brazil could trigger a hurricane in Texas. In fact, a continuum exists that connects the atmosphere and all correlated phenomena all over the entire planet. Consequently, the heating of the ocean at a particular point in the middle of the Atlantic, for example, could alter a depression in North Europe. Shifting a hurricane away from the area of New Orleans would prevent a great loss of life and damage there, but could cause a change in the behavior of storms in Mexico, the Caribbean or South America, which could be catastrophic for local agriculture.

One nation controlling their climate could inadvertently influence and control the climate of other nations. All this could unleash arguments and never-ending conflicts. It is even more alarming to consider what could happen if the military uses this technology. In fact, diverting the path of a hurricane in a preventive way towards a safer area would mean saving a lot of people and property in the area that would otherwise be hit by the hurricane. But if this is carried out with military aims, it could mean to fire a really powerful weapon on a given target, with immense damage for the whole area.

In a study which is periodically done by the *US Air Force* to identify strategic weapons of the future, the control of the weather is considered as one of the most important.

In general, the change of atmospheric conditions can favor planned military actions. In some cases, the change of the weather can also be used as a vehicle for the diffusion of toxic substances. In the last year of the First World War, tens of thousands of soldiers died due to an aggressive chemical that was released into the atmosphere, because a sudden change in wind direction moved it towards their own troops instead of the enemy.

What would have happened if the winds had directed the toxic substance to an area inhabited by civilian populations? Therefore, the use of aggressive biological and chemical warfare was prohibited by the Geneva Convention in 1925. Unfortunately, the convention was limited by not prohibiting the development, production and storage of these substances. In fact, the production of biological weapons does not require a massive investment. As a consequence, even nations with a small military strength could produce and use these aggressive agents in a war.

During the United Nation's environmental conference, held in Stockholm in 1972, it was established that the atmosphere belongs to everyone and must not be considered unilaterally by any one country, without taking into account the interests of other nations. On the basis of this declaration, and following the possible threat of a meteorological war, in December 1976, more than 50 Countries, among them the majority of the European countries, the Unites States, Russia, Australia, India and Brazil, signed a convention that prohibited any attempt aimed at modifying the climate for military aims[8]. Paragraph 1 of article I of the convention states the following:

> *"Each State Party to this Convention undertakes not to engage in military or any other hostile use of environmental modification techniques* having widespread, long-lasting or severe effects as the means of destruction, damage or injury *to any other State Party."* (italic emphasis added, RM)

Substantially, the agreement prohibited the use of any methods or technologies which could harm the ecosystem of a determined region for long periods of time. Even though, according to the convention, these technologies could be applied in cases where the effects lasted for a few months and would not provoke disturbance or serious damage to human life, natural resources and economies. As well, independently of this rule, they can be applied to limited areas, defined in the text of the convention as *more than hundreds of square kilometers.*

Article II of the Convention clarifies, instead, which are the techniques that could effectively influence the natural processes. It declares the following:

> *"As used in article I, the term 'environmental modification techniques' refers to any technique for changing – through the deliberate manipulation of natural processes – the dynamics, composition or structure of the Earth, including its biota, lithosphere, hydrosphere and atmosphere, or of outer space."*

The natural phenomena of which the Convention speaks could be earthquakes, tsunami, changes of the ecological equilibrium of a region, alteration of atmospheric conditions (clouds, precipitation, different types of cyclones, tornados), alterations of climatic conditions, the currents of the oceans, the ozone layer or the ionosphere layer.

Unfortunately, it is to be noted that these techniques could easily be used by states that did not sign the Convention. In this case, the influence on the atmosphere can have remote effects not limited to the borders of the target country, but can affect many others, even neutral areas. It is therefore obvious that this treaty is a great step towards a total disarmament, but does not completely solve the problem and does not eliminate all the dangers.

A global cooperation between all nations is necessary to guarantee a total security in the use of these techniques, like the Russian Federov, an expert in techniques to modify the climate, confirms[9]:

> *"The problem of global and regional climate modification is essentially international and requires the common efforts and coordinated planning of all nations."*

In 1966, McDonald, head of the commission on the study of the results and the implications of the modification of the climate, financed by the *American Academy of Sciences,* wrote the following about the use of these techniques in his final report[10]:

> *"One could, for example, imagine a field commander calling for local enhancement of precipitation to cover or impede various ground operations. An alternative use of cloudseeding might be applied strategically. We are presently uncertain about the effect of seeding on precipitation downwind from the seeded clouds, but continued seeding over a long stretch of dry land clearly could remove sufficient moisture to prevent rain a thousand miles downwind. This extended effect leads to the possibility of covertly removing moisture from the atmosphere so that a nation dependent on water vapor crossing a competitor country could be subjected to years of drought. The operation could be concealed by the statistical irregularity of the atmosphere. A nation possessing superior technology in environmental manipulation could damage an adversary without revealing its intent."*

In military circles, weather modification is not considered to be of high strategic importance at the moment, but there is already talk about a possible reevaluation of the decision made by the United Nations. Therefore, weather and climate could occupy a primary position in a future world war.

The first military attempts of intervention to create prevailing unfavorable atmospheric conditions for the enemy in a conflict dates back to the start of the last century, during the First World War. At the end of 1916, both the Italians and the Austrians intentionally caused a long series of avalanches in the direction of the enemy troops. The result was the death of more than 15,000 soldiers. A case of accidental modification of atmospheric conditions occurred during the bombardment of London, Coventry, Hamburg and Dresden in World War II, where a number of fires were ignited by phosphorus bombs.

The rising air currents, heated by the fire, favored the formation of strong winds in the surrounding area which reached a speed of about 200 km/h, causing destruction and death everywhere. This phenomenon was called *firestorm*.

The experience gained from these incidents was used by the Americans to intentionally cause strong destructive fires and winds two different times, in 1961 and 1966, in Vietnam.[11] Anyhow, in both cases the results were not those hoped for, because the humidity and climatic conditions of the jungle were unfavorable, impeding the development of strong fires and winds, and preventing a human tragedy as well as an ecological disaster.

In 1967, still during the Vietnam War, the Americans put into act operation *Popeye*[12]. The goal of the project, carried out in total secrecy, was to increase rainfall along the border regions of Laos, Cambodia and North Vietnam. They hoped swamp and block the Ho Chi Minh Trail, a North Vietnamese transport route snaking through hundreds of miles of dense foliage into South Vietnam. The objectives were to increase the precipitation at the start and end of the monsoon season. The abundant rain would have flooded the supply routes and turned the roads into mud.

The operation was carried out through the chemical fertilization of clouds with silver iodide over particular strategic areas during more than 2,500 aerial missions. The results were conflicting. There is no specific data on the increase of precipitation, since the only parameter monitored during the whole operation was the density of traffic on the Ho Chi Minh Trail, measured through automatic sensors. However, the effectiveness of the methods used is in doubt, and no one knows with certainty if the abundant rain that hit the area was the result of the military operation or the natural development of the local climate, or both. The project was immediately suspended when an American journalist became aware of it and published details of the operation in the press.

Based upon such considerations, in 1974 Mexico went so far as to accuse the United States of intentionally diverting the hurricane *Fifi* from Florida to Central America, causing the death of thousands of people. However, no evidence to support this accusation was provided. At that time, no technique accepted to mainstream science existed which could securely change the path of a hurricane.

But in which way could all these interventions to modify the atmospheric and climatic conditions, real or otherwise, be carried out?

One of the most widespread methods for stimulating rain clouds is *cloudseeding*. It is based on the principle of artificial nucleation and ice condensation at low temperature in cloud. Each seeded crystal artificially introduced into the cloud functions as a catalyst for the aggregation of ice or small drops, until snow crystals or raindrops have formed that are heavy enough to fall to the ground.

Not all clouds are suitable to be seeded. It is necessary that they possess a characteristic similar to the clouds that can make rain under natural conditions, such as vertically-developed cumulonimbus.

One of the biggest problems one faces in this type of research is the objective evaluation of the results, or the impossibility of proving that the method or the procedure really has worked.

In fact, one cannot concretely attest to how much rain would have fallen if the behavior of the clouds had not been altered, because it is not possible to know what would have happened in the absence of seeding.

Therefore, a great difficulty exist in evaluating and objectifying the obtained results. Everytime, the verifiability is missing, as it is the case with all non-repeatable experiments on natural phenomena. In these cases, the only way of evaluating the results is to use statistical methods.

The first research began in the 1930s when scientists discovered that raindrops form more quickly when ice crystals are present in the clouds.

During those years, the Dutchman Veraart seeded the clouds with dry ice and supercooled water, which remains in a liquid state even at temperatures lower than the freezing point (for water, this is 0°C).

The first large scale experiments reported in the literature began in November 1946. Schaefer[13] and

Langmuir[14] carried out tests by dropping 3 pounds of very fine dry ice from an Aeroplane, above a large layer of clouds at very low temperatures. Within five minutes, the formation of snowflakes occurred which fell for around 1,000 meters before being transformed into rain.

A big hole formed in the cloud, directly corresponding to the fertilized area. Obviously, the results of the experiment attracted the interest of the government and farmers, in constant battle against drought, and also the military.

A short time later, Vonnegut[15] replaced dry ice with silver iodide, which has a similar crystalline structure, but is much cheaper.

Following this discovery, Langmuir undertook a series of experiments during the 1950s that involved fertilizing clouds with silver iodide[16] once-per-week in New Mexico. He noted an interaction between the various atmospheric systems even in distant areas, more than 1,000 km away, with the formation of storm systems and abundant rain showing a regularity of 7 days, in northeastern New Mexico, in the area of Mississippi, and in the Ohio River Valley.

A series of experiments were then performed, both in the laboratory and on a larger scale, aimed at validating the theory which had been developed until then. Tests were carried out on different types of clouds using different activation substances and techniques, and under a great number of different atmospheric conditions.

The obtained results were always conflicting. In some cases, effects were registered which were difficult to explain, such as the dissolving of the clouds and the reduction in precipitation in Central Mexico. In other cases, as in Australia and the United States, the effects were positive and exceeded all expectations.

The *cloudseeding* operations carried out in the United States from 1972[17] to 1973[18] (on a budget of around 50 million dollars) registered an increase in the precipitation from 5 to 30%[19], in some isolated cases up to 50%[20]. Apart from these results, certainly positive, the meteorologists noted the presence of secondary effects, with negative consequences on the climate. Among them, a reduction of rainfall in the periods following the seeding stood out. In addition, it has been ascertained that in some cases, the artificial seeding of clouds could also cause contrary effects such as the inhibition of the natural formation of rain. In the report from the *US Department of Commerce*[21], the following can be read:

> *"... Earlier work had shown that a deterioration occurs in the effectiveness of successive seeding experiments. Seeding was conducted in alternate years to avoid possible carry-over effects from the persistence of the seeding agent in the seeded area ... a possibility exists that rainfall was reduced during the summer season in one portion of the seeded area ... Rainfall observed in the target area had an overall decrease compared to the control area. Thermodynamic effects related to the seeding are suspected of causing this variation ... However, seeding does not always produce precipitation and there is evidence that, under some conditions, it may actually hinder precipitation ..."*

The report of the *US Department of Commerce* on the interventions carried out by the United States in 1974[22], shows that in that year numerous *cloudseeding* operations were performed. A large number were carried out in Texas (12), California (11), and Washington (7), and many others in the region surrounding the Great Lakes. In the following years, between 1975 and 1977, a long period of severe drought was observed in the areas with the highest density of interventions. It is therefore possible that the massive emission of chemical agents into the atmosphere might contribute to the formation of undesirable climatic effects.

In the 1980s, new research developed which has revolved around the deflection and the destruction of hurricanes. Hurricanes, also called typhoons, have an enormous destructive power and cause immense damage in areas which they strike. The most risky seasons are summer and autumn, when the seas to the north and south of the equator are heated. Australia, the Philippines, China, Japan, the Caribbean Islands, Mexico, the east coast of the USA and the west coast of Africa are at greatest risk.

From the experimental *Project Stormfury*[23] it appeared possible that the maximum velocity of a hurricane might be reduced by chemically fertilizing its cloud-bands with silver iodide, in the hope this would deprive the core-eye of the hurricane of moisture necessary to maintain its strength. The seeding is done with particular aircrafts, called *hurricane hunters*, which are able to survive in the eye of a typhoon. Many experiments were undertaken in this direction, with encouraging but inconclusive results. International concerns forced an end to the experiments.

One sector particularly interested in the application of artificial methods for the modification of atmospheric conditions is the tourist sector. In these cases, it is about the production of artificial snow in areas with ski slopes. The first studies aimed at creating artificial snow were developed thanks to a chance discovery in Florida at the end of the 1930s. Following a quick drop in temperature, an automatic irrigation system sprayed snow instead of water. Having discovered the principle, it was much easier to apply it to covering of ski slopes. Starting from the 1960s, snow cannons spread through the mountains both in the United States (New England, the Rocky Mountains) and in Europe (the Alps). Contrary to *cloudseeding*, the artificial snowmaking systems use totally natural elements and conditions: water and air under specific temperatures, humidity and pressure conditions. Spraying air and water, finely pulverized at low temperatures, allows the water to crystallize immediately into very fine granules.

A branch of the modification of the climate that is really useful especially in agriculture, is devoted to anti-hail systems. Some procedures, called *active methods*, act directly on the hailstones where they are forming. To obtain a larger number of smaller hailstones, the clouds are seeded with silver iodide. These are then transformed into rain during their descent to the ground through friction with the air. In general, airplanes carry out the seeding both above and below the clouds. Good results have been achieved also through the seeding above the clouds using dry ice.

Instead, *passive methods* include the launching of plastic rockets that contain explosive trinitroluene (TNT). These are shot and exploded among the falling hailstones, at a maximum height of about 2 km. The blast is intended to disintegrate the stones prior to their impact with the ground.

Another method is to fire blank shots from cannons which form airwaves that spread from the earth's surface upwards to an altitude of some hundred meters. Also in this case, the waves affect the formation and falling mechanisms of the hailstones.

Chapter 3 Citations

1. Aristotle, *Meteorologia*, Bompiani, Milan, September 2003.
2. Jean Baptiste Joseph Fourier (1768-1830), French mathematician. He developed a method of mathematical analysis in which the periodic function can be decomposed in elementary periodic curves. He also developed a series of partial differential equations that govern the diffusion of heat. Furthermore, he was responsible for a number of engineering projects, among them the construction of the road connecting Turin to Grenoble.
3. Svante August Arrhenius (1859-1951), Swedish chemist. Studied the conductivity of ionic solutions and developed a law that describes the rates at which chemical reactions occur. He received the Nobel Prize for chemistry in 1903.

4. Vilhelm Frimann Koren Bjerknes (1862-1951), Norwegian physicist and pioneer of modern meteorology. He applied the hydrodynamic and thermodynamic laws to the flow of the air masses in the atmosphere, in order to predict the meteorological conditions of a given area. His theories and his work on meteorology contributed to the subsequent development of the wireless telegraph.

5. Tor Harold Percival Bergeron (1891-1977), Swedish meteorologist, known for his studies on the physics of clouds and the formation of warm and cold fronts. He was the first meteorologist to consider the influence of the upper atmosphere on weather formation close to the earth's surface. He closely collaborated with Bjerknes on the development of methods for weather forecasting.

6. Halvor Solberg (1897-1975), Norwegian meteorologist, was known for his studies on air masses and the discovery of the polar front.

7. Mullins, J.: *Raising a Storm,* New Scientist, Vol. 175, No. 2353, 27 July 2002.

8. Organization of the United Nations, *Convention on the Prohibition of Military or any Hostile Use of Environmental Modification Techniques*. The Convention ended in New York on December 10th, 1976. It was signed on the 18th of May 1977 and came into effect in most of the nations on the 5th of October 1978.

9. Haas, J.E.: *Weather and Climate Modification*, WH Hess Editor, New York, 1974.

10. *Weather and Climate Modification: Problems and Progress*, National Academy of Sciences, Committee on Atmospheric Sciences, National Research Council, Washington DC, 1966.

11. Caroselli, G.: *Il Tempo per Tutti*, Mursia Editore, Milan, 1995.

12. Breuer, G.: *Weather Modification*, Cambridge University Press, Cambridge, 1979.

13. Vincent J. Schaefer was born at Schenectady, New York, in 1906. In 1926, he started to work at the General Electric Laboratory as a mechanic. In 1932 he became Langmuir's assistant and was primarily engaged in research regarding military applications. From 1959 to 1961, he was the director of the *Atmospheric Science Center* at *Loomis School* in Connecticut. In 1966, he became director of the *Atmospheric Sciences Research Center* at the University of New York. He died in 1993.

14. Irving Langmuir was born in Brooklyn, New York, in 1881. He graduated in metallurgical engineering from *Columbia University* in 1903 and received his Ph.D. from Gottingen University in 1906. In 1909, he started to work at the General Electric Laboratory in New York. He studied the theory of atomic structure and molecular films, and thereby opened new horizons in the biochemistry and chemistry of colloids. He was also interested in the phenomena related to the electrical discharges in gasses. In 1932, he received the Nobel prize for chemistry. He died in 1957.

15. Bernard Vonnegut was born in Indianapolis, Indiana, in 1914. He graduated in Chemistry from MIT in 1936 and received his Ph.D. in 1939. After a short period as a researcher at MIT, he moved to General Electric in 1945, where he was occupied with cloudseeding. From 1952 on, he worked under Arthur D. Little, where he studied problems related to atmospheric electricity. Since 1967, he was a dedicated teacher at the department of Atmospheric Sciences at the University of New York. He died in Albany, New York, in 1997.

16. Langmuir, I.: *Analysis of the Effects of Periodic Seeding of the Atmosphere with Silver Iodide*, in *The Collected Works of Irving Langmuir, Vol. 11, Cloud Nucleation*, Pergamon Press, New York, 1961.

17. *Summary Report: Weather Modification*, FY 1972, US Dept. of Commerce in Eden, J.: *Atmospheric Murder. The Cause of Global Weather Chaos*, Journal of Orgonomy, Vol. 12, No. 2, Orgonomic Publications Inc, New York, November 1978.

18. *Summary Report: Weather Modification*, FY 1973, US Dept. of Commerce, in Eden, J.: *Atmospheric Murder. The Cause of Global Weather Chaos*, Journal of Orgonomy, Vol. 12, No. 2, Orgonomic Publications Inc, New York, November 1978.

19. *What Can We Do About the Weather?,* Pacific Northwest Regional Commission, 1977, in Eden, J.: *Atmospheric Murder. The Cause of Global Weather Chaos*, Journal of Orgonomy, Vol. 12, No. 2, Orgonomic Publications Inc, New York, November 1978.

20. *Summary Report: Weather Modification*, FY 1972, US Dept. of Commerce in Eden, J.: *Atmospheric Murder. The Cause of Global Weather Chaos*, Journal of Orgonomy, Vol. 12, No. 2, Orgonomic Publications Inc, New York, November 1978.

21. *Summary Report: Weather Modification*, 1969-1971, US Dept. of Commerce in Eden, J.: *Atmospheric Murder. The Cause of Global Weather Chaos*, Journal of Orgonomy, Vol. 12, No. 2, Orgonomic Publications Inc, New York, November 1978.

22. *Weather Modification Activity Reports*, CY 1974, US Dept. of Commerce, 1975, in Eden, J.: *Atmospheric Murder. The Cause of Global Weather Chaos*, Journal of Orgonomy, Vol. 12, No. 2, Orgonomic Publications Inc, New York, November 1978.

23. *Weather and Climate Modification: Problems and Progress*, National Academy of Sciences, Committee on Atmospheric Sciences, National Research Council, Washington DC, 1973. p.106-108.

Cosmic Orgone Engineering

> *"... He claims Moses simulated his miracles. He claims Moses invented the Ten Commandments. He maintains the Bible is a book of fables. He believes that all, even the demons, will save themselves ...."*
>
> Roman Inquisition on Giordano Bruno

Cosmic Orgone Engineering (CORE) was conceived by Wilhelm Reich at the start of the 1950s. It was based on the principle that the Earth's atmosphere, as well as the cosmos in general, is not empty but is filled with orgone[1], the same form of energy that Reich discovered in 1938 while studying bions. It should be pointed out, however, that this assumption, fundamental for the development of orgonomy, has not yet been accepted by current conventional physics, which largely bases its convictions on the principles of Einstein who postulated an empty universe devoid of energy.

CORE is a sector of orgonomy that deals with the activities and techniques aimed at varying the concentration of orgone energy present in the atmosphere, both under dynamic and static (DOR[2]) conditions. In particular, it is possible through the use of appropriate instruments and procedures, to influence the conditions and the concentrations of atmospheric orgone, which is responsible for the meteorological phenomena and climatic conditions.

It is essentially possible through intervention to beneficially influence, or negatively disrupt, the atmospheric phenomena which characterize the climate of a certain area, and which are dependent, to a greater or lesser extent, on the interactions of orgone energy, oxygen, water vapor, rain, sun, and wind.

This culminates in a powerful method to restore natural rains, to influence winds and lower the rate of atmospheric pollution, to eliminate fog and affect humidity, and to divert hurricanes.

Reich conceived of this technique as a method of reestablishing natural atmospheric functions and rains within dry, arid or desert[3] areas, something which requires a spontaneous motion of orgone energy with natural charge and discharge phases, and periodic cycles of precipitation and clear weather.

CORE is different in substantial ways to traditional weather modification, which was basically developed to forcibly modify or manipulate atmospheric conditions in our favor, or even for military purposes, without ever taking care about natural atmospheric behavior.

CORE also presents numerous advantages over traditional methods. It does not require chemical substances to bring about changes in meteorological conditions. The technique can be applied under nearly any conditions, which is not possible with *cloudseeding*, as that always needs a cloud coverage with certain preexisting characteristics. In addition, it is very cheap when compared to the high costs of the traditional methods.

In the development of this technique, Reich drew upon his experiences gained from the character analysis and vegetotherapy of human beings. With those therapeutic methods, he tried to eliminate the energetic blocks of the individual, located in the parts of the body where muscular armoring was present, thus restoring the natural functioning of the organism by reestablishing the natural energetic flow. In parallel, he recognized that the arid and desert areas of the planet, which did not experience regular rainfall, were subjected to blocks in the orgone energy flow comparable to what happens in the organism. These areas were characterized by an increase in *environmental illness*es such as drought, aridity and desertification.

The intervention with particular CORE techniques in these areas suffering from atmospheric stagnation, are intended to eliminate the blocking and to reestablish the natural orgone energy flow and pulsation in the atmosphere, thereby restoring humidity, rain and natural climatic conditions.

Orgone Physics

Cosmic Orgone Engineering is essentially based on the laws that govern the behavior of orgone energy, as well as on certain properties that metals or particular materials and substances possess in regards to orgone.

One of these is the law of the orgonomic potential which regulates the energy flow that moves from one system to another. It states that the orgone energy always flows from the lower to the higher energy potential, as Reich described in his article *DOR Removal, Cloud-busting, Fog-lifting*[4]:

> "The 'orgonomic potential' denotes all functions in nature which depend on the flow of cosmic energy, or potential, from LOW TO HIGH or from WEAKER TO STRONGER SYSTEMS. …. The orgonomic potential is most clearly expressed in the maintenance in most animals on this planet of a temperature higher than that of the environment, and in the function of *gravitational attraction*. In both cases, the stronger energy system draws energy from or attracts a weaker system nearby…"

Orgone is easily conducted by metals and has a high affinity to water, which it attracts, absorbs and holds.

On the basis of these simple principles, Reich created a real and actual science of the atmosphere within orgonomy. The interventions which Reich carried out on the atmosphere were performed using an apparatus called the *cloudbuster*, while the activity was called *cloudbusting*.

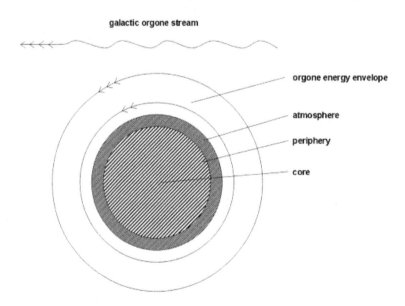

galactic orgone stream

orgone energy envelope

atmosphere

periphery

core

Figure 2: Orgone Envelope and Galactic Orgone Streams (from Reich[6])

One of the most important discoveries which allowed Reich to apply and develop *cloudbusting,* was the identification of an *envelope* of orgone energy surrounding the Earth[5]. He recognized that this orgone envelope moves from west to east, with a velocity greater than the Earth's rotation.

This envelope not only covers the whole surface of the globe with a thickness of thousands of meters, but also penetrates deeply into the Earth's crust towards the center of the planet. Furthermore, it extends beyond the atmosphere out into space, merging with the Equatorial and Galactic orgone streams which exist beyond the immediate gravitational field of the planet. Reich observed these two energy streams intersected at an angle of 62 degrees, the Equatorial Stream being aligned with the Earth's equatorial plane, and the Galactic Stream moving from southwest to northeast, aligned with the plane of the Milky Way Galaxy. In addition, these energy streams move and support the Earth in its translation and rotation through the cosmos, somewhat like a ping-pong ball that floats on the surface of the sea and is moved ahead by the sea-waves.

Figure 2 shows a graphic representation of the orgone envelope surrounding the globe, and the Galactic Orgone Stream.

Reich compared the energetic expansion and contraction of the orgone envelope to a living organism that spontaneously pulsates in accordance with the seasons and the atmospheric conditions. This behavior is essentially local and changes from area to area, according to the geographical location.

In spring and especially in summer, the orgone envelope expands towards space with a force which opposes the force of gravity. This tendency is more noticeable in hot regions that are situated at low latitudes, while it is less apparent in cold regions at higher latitudes. In autumn however, the envelope starts to contract towards the core of the planet, thinning more and more, until it reaches its lowest level in winter.

All this happens also with variations in meteorological conditions and, therefore, with the alternation of sunny and rainy weather. During sunny weather, the envelope expands outwards into space, while when rainy weather comes along, it contracts and concentrates at the Earth's surface.

This behavior can be observed anywhere in nature and explains a large part of natural phenomena. For example in the countryside, the farmer predicts sunny weather when the smoke leaves the chimney in an upward motion into the atmosphere[6]. However, when rainy weather approaches, the smoke tends to move horizontally and parallel to the surface of the Earth. According to Reich, during the sunny weather, the smoke of the chimney is able to rise upwards due to the expansive force of the orgone envelope, which creates a force strong enough to oppose the force of gravity. The opposite happens when rainy weather arrives. In this case, the contraction of the orgone envelope causes a reduction in the expansive force that balances the force of gravity. So while the force of gravity always prevails, the smoke is not able to easily rise and move vertically into the atmosphere, as it does during sunny weather.

The same principle applies to the formation of autumn morning fog. During the summer days, fog rarely exists in the valleys, while in autumn it develops even in valleys situated at a low altitude. This is because, in summer and in warm areas situated at low altitudes, the expansive and the dispersive force of the orgone envelope opposes the force of gravity. The water vapor rises upwards, dispersing into the atmosphere like the smoke from the chimney. In this case, it is difficult for fog to form. In autumn and winter, particularly in cold regions at high altitude, the expansive force of the orgone envelope is greatly reduced, and therefore the water vapor is no longer able to rise and disperse, instead accumulating in higher concentrations closer to the ground, resulting in the formation of fog. At sunrise, the weakened and contracted orgone energy experiences an excitation and an increase in its expansive force. The water vapor is therefore able to rise and disperse into the atmosphere.

Another typical example of the pulsatory nature of the orgone envelope and the law on the orgonomic potential, are the climatic behavior characteristics of a certain region. Consider for example desert areas, where the air is always dry and it never rains, while in other areas there is much humidity and rain for weeks on end, or even areas where there is much humidity but little rain. In all these cases, both the distribution and the dynamic conditions of the orgone energy in the atmosphere play a role. Where it is equally distributed and present in large quantities, like in the summertime Mediterranean dry period, the water vapor in the atmosphere is sufficiently uniform and regular, and is continually being dispersed. Large concentrations of energy in any one place do not exist, and the expansive and dispersive force of the orgone prevails, only occasionally giving rise to the formation of clouds and rain. However, in mountainous areas, where a great difference in the orgone concentration exists, there is much rain even if there is little humidity in the air. This is partly because of the difference of the orgonomic potential present in the atmosphere that continually causes concentrations of water vapor and the formation of clouds.

Another natural phenomenon which often intrigues for its incomprehensibility, is the formation of a ring of varying diameters around the Moon or the Sun, prior to the arrival of rainy weather. The significance of this phenomenon has been well noted in agricultural circles, and many farming activities are planned on the basis of it, as it is an announcement of stormy weather and rain.

It is known that the orgone envelope contracts prior to rain. This creates an increased concentration of orgone in the atmosphere, favoring the accumulation and suspension of water vapor particles, which then tends to lead to the formation of rain-bringing clouds. This also creates thin layers of high altitude refractive ice crystals, which in turn makes an observable ring around the Sun or Moon. However, the appearance and size of the ring, well visible to the naked eye, depends solely on the level of contraction of the orgone envelope that covers the surface of the Earth. It is visualized by the rays of the Sun or the Moon that hit the external surface of the envelope and can easily be seen from the Earth.

Figure 3 schematizes the variation of the ring with the change in the contraction of the orgone envelope. Upon contraction, the envelope creates an ellipsoidal surface with the rays of the Sun, or the Moon, that hit and penetrate the orgone envelope.

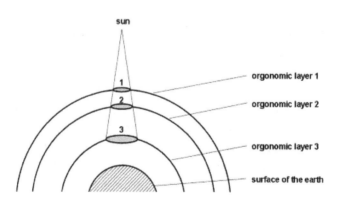

Figure 3: Contraction and Expansion of Earth's Orgone Envelope (from Reich[6])

In the first orgonomic layer, the envelope is sufficiently expanded, and the ring is very small and difficult to see as the sun hides it. Gradually, the envelope contracts and the weather turns towards wet conditions, increasing the surface that the rays of the Sun (or the Moon) form with the orgone envelope. In the third instance, the orgone envelope is in the most contracted state, and the ring that forms is the most extended and the most visible to the naked eye. Consequently, the greater the contraction of the orgone envelope covering the Earth's surface, the wider the ring will be that forms around the Sun or the Moon, and vice versa.

Finally, one of the most significant examples for the pulsatory nature of the orgone envelope is the way in which nature organizes itself according to the variation of the seasons. Reich saw that the increase in the temperature and the heat in spring and summer are indicators of expansion and dispersion of the orgone envelope outwards into space. On the other hand, the reduction of temperature and the cold of autumn and winter, are the result of the increasing contraction of the envelope towards the center of the globe. Heat and cold are therefore connected to the terrestrial orgone envelope's activities of expansion and contraction.

Just before the start of spring, a general increase in the activities of nature is noticeable. The trees reawaken, the stiffness disappears, and they develop a great liveliness in the branches. It is possible to observe an increase of the energetic expansion and activity levels, of the rise of sap in trees and other plants, with greater aliveness in every living thing.

It is the flow of orgone energy, due to the expanded orgone envelope, that allows the sap to rise in the plants, defying the force of gravity. The chlorophyll increases quickly in the trees. The superimposition of the yellow (color of the resin) and blue (color of the orgone) is the origin of the green color of the leaves. The resin slowly transforms into green, thanks to the blue of the orgone absorbed from the ground and the atmosphere. In addition, the appearance of the first buds is a clear expression of the expansive force of the orgone energy. It is the orgone energy that pushes against the rigid outer membranes, sufficiently elastic at a young age, to cause them to expand, develop and grow. The enormous potential of this force can easily be observed in the blade of grass that pushes through the soil in order to grow and develop.

In autumn instead, the contraction of the orgone envelope towards the center of the planet causes living nature to become depleted and dry up. The leaves turn yellow through the loss of the blue component (or the orgone energy) from the green color, they dry up and detach from the branches due to the loss of cohesive energy. The increased activity of the sap, so typical for spring, disappears. The temperature and water vapor in the air behaves in the same manner and transforms into snow, crystallized water vapor containing orgone energy.

The Cloudbusting Technique

Reich first applied *cloudbusting* during an emergency which hit Orgonon between the months of March and September 1952. During this time, a great concentration of dark-colored DOR clouds was situated over the area.

At first, Reich thought that the DOR clouds, coming mainly from the west, could be the result of an nuclear test carried out in the western part of the United States. After having determined that none had been carried out in that time, he concluded they were the product of the *Oranur Experiment* he was undertaking, as described in Chapter 2.

When the illness brought by the DOR clouds became intolerable by the end of April 1952, he tried to remove them. He remembered a casual and apparently insignificant observation he had made many years before. In 1940, he had observed by chance that pointing some long metallic tubes toward the surface of Lake Mooselookmeguntic at Rangeley, Maine, seemed to influence the wave movement of the lake's water. This occurrence had appeared very strange to him. Based on this experience, he assembled a number of metal tubes between 2.70 and 3.60 meters long, with a diameter of 3.7 cm, mounted them on a wooden support, connected them via flexible metal cables to a deep well and pointed them at the blackish cloud concentrations.

The effect he obtained was immediate. The DOR clouds started to contract and dissipate. And when the tubes were pointed to the western horizon, after a few minutes, a breeze started to blow from west to east. He therefore concluded that with this procedure, he was able to bring in fresh amounts of orgone energy where stagnant and noxious clouds had predominated only a short time before. Soon he discovered that even storm clouds could be influenced by using these tubes in a well defined way, increasing or decreasing them, and altering the patterns up to an area of several thousands of square kilometers. He also noted there was a delay, sometimes even of several days, between the start of the experiment and the effects. All of this was contrary to what happens with *cloudseeding*, which influences only individual clouds, in local regions only, and has rather quick influences.

Thus, CORE and *cloudbusting* were born. In the autumn of 1952, Reich constructed in Portland, Maine, the first apparatus intended for operations of Cosmic Orgone Engineering, which became known as the *cloudbuster*. In the following months, he began to undertake experiments which became more and more systematic and successful. He developed increasingly effective protocols for the production or prevention of rain.

Between 1952 and 1957, he carried out numerous operations in the United States (photos 1 and 2), most of them aimed at bringing rain to areas suffering from drought. He published the majority of his results in the *CORE* journal.

Many of these operations, which he referenced using the term *OROP*[7], took place in Maine[8,9,10,11,12], primarily in the area surrounding Rangeley. Others were performed near different cities along the Atlantic Coast[13], such as New York, Philadelphia, Washington, and Savannah. His most important and long-term experiment was carried out near Tucson, in the desert of Arizona[14,15,16].

On one occasion, he intervened to weaken hurricane Edna, which was threatening the coast of Maine, and successfully deflected its course out towards the Atlantic Ocean[17]. In a final case, he carried out an operation to delay the rain that was brewing over the whole area of Rangeley, to help realize a parade organized for the children of the area[18].

He undertook other interventions to produce rain at the request of local farmers. Merrill, the owner of the *Blueberry Factory* at Rangeley, remembers his experiences with the *cloudbuster* used to bring rain for the crops that otherwise would have been lost due to a shortage of water[19]:

> *"He said he'd bring us an inch of rain for a thousand of dollars, and I said, "Well worth it!". And the next morning he called, and he said Mr. Reich wouldn't guarantee an inch but he would guarantee rain. So we said, "Go ahead just the same". So, on the Monday morning, still was no rain, and he came through from Rangeley with his equipment and set it up at Graham lake. It was here, and it was on a truck, a big truck. He set it right here and ran something into the water, I suppose to ground it. And he picked up at noon, and left, and it was still clear, but these funny clouds were going. Then, that evening or that night around midnight, we got a quarter of an inch of rain all through this section within a radius of probably 25 miles."*

In the majority of the experiments, Reich obtained the intended results. These were evaluated on the basis of the official weather forecast for the day of the experiment. To strengthen the obtained results, an operation was only scheduled when the weather forecast had predicted no rainfall for at least two successive days. In more complex operations, such as those fighting droughts along the Atlantic coast and in the desert of Arizona, or to divert the course of hurricane Edna, the official precipitation data supplied by the US Weather Bureau was consulted.

The *cloudbuster* is essentially a tool capable of absorbing orgone energy from the atmosphere, whatever its energy-potential may be. The affected zone where it is aimed gradually loses orgone energy in favor of the surrounding areas, thereby modifying the original orgone concentrations.

Eden briefly explains the functioning of the instrument at the beginning of his book *Core Manual*:[20]

**Photo 1 – Wilhelm Reich During a Cloudbusting Operation in the mid 1950s
(Courtesy of the Wilhelm Reich Museum)**

"There is no secret concerning the basic principle ruling the functions of the Cloudbuster (CLB). Orgone energy flows from the weaker to the stronger energy system. The simplest CLB is a piece of metal pipe that is grounded into water. Atmospheric energy will move from the atmosphere, through the pipe, and into the water, the water being the stronger energy system. This Reich called "orgonomic potential", which is exactly opposite to the way that secondary-energy systems move."

In conceiving the *cloudbuster,* Reich was inspired by the functioning principle of the lightning rod. The tip of the lightning rod attracts the lightning and discharges it into the Earth through metal cables. To Reich, lightning was nothing else than a discharge of atmospheric orgone energy that occurs within a very short interval of time and in an extremely narrow space. With the discharge, it follows the orgonomic principle of orgone flow from the weaker to the stronger energy system, which in this case is the Earth's surface.

Yet the *cloudbuster* is different from a lightning rod in some fundamental ways. The *cloudbuster's* purpose is not to attract lightning and discharge it to the Earth, but to gradually affect the orgone energy concentrations in the atmosphere or in the clouds. Orgone energy charge is absorbed slowly, in small quantities, and not in the form of a sudden discharge.

In addition, the *cloudbuster* consists of long hollow tubes, not of solid steel like the lightning rods. The tubes, of a specific number and length, have the task of absorbing and carrying along the orgone energy stream coming from the atmosphere towards the surface of the Earth. The orgone stream is generally directed into water, and not to Earth as with the lightning rod. The water must be flowing water, like in creeks, rivers and non-stagnant lakes. Water is the preferred discharge medium, because the attraction between water and orgone energy is greater than between orgone and many other elements or substances present in nature.

Reich developed very simple and practical protocols both for the creation and the dissipation of clouds. In case he wanted to create a cloud, and thus to stimulate rain, Reich employed the following operational rules[21]:

"... we draw ENERGY from THE CLOSE VICINITY OF THE CLOUD IF WE WISH TO ENLARGE EXISTENT CLOUDS AND TO PROCEED TOWARDS RAIN-MAK-ING ...."

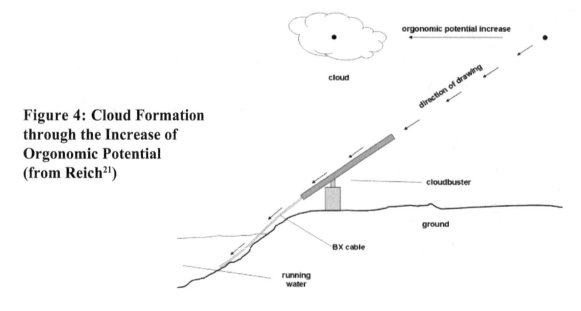

Figure 4: Cloud Formation through the Increase of Orgonomic Potential (from Reich[21])

To enlarge a cloud, energy is absorbed from an area immediately surrounding it, lowering the orgone energy charge at the absorption point. Consequently, a flow of orgone is created from that point towards the cloud, which possesses a greater energy charge. This will increase the energy of that cloud, until its maximum energy level possible is reached, or until the saturation of the cloud, after which there is rain. Figure 4 summarizes the absorption technique developed by Reich to make rain.

By disrupting the regularity of the orgone energy distribution in the atmosphere, it is even possible to create new clouds in a clear sky. In this case, one starts with special methods to mobilize and excite the energy in the atmosphere. However, a big difference exists between getting rain from a cloudy sky and from an initially cloudless sky. In fact, the more clouds are present in the sky, the easier it is to create rain. On the other hand, the fewer number of clouds present in the sky, the more difficult it becomes.

Furthermore, it is possible to dissolve clouds present in the atmosphere by withdrawing orgone energy from their center. In this event, the cohesive energy that holds the water vapor together is lowered and the clouds tend to dissolve, as Reich himself reported in CORE[22]:

"ONE DISSIPATES CLOUDS OF WATER VAPOR BY WITHDRAWING, AC-CORDING TO THE ORGONOMIC POTENTIAL, ATMOSPHERIC (COSMIC) OR ENERGY FROM THE CENTER OF THE CLOUD. THIS WEAKENS THE COHESIVE POWER OF THE CLOUD. THERE WILL BE LESS ENERGY TO CARRY THE WATER VAPORS, AND THE CLOUDS NECESSARILY MUST DISSIPATE. THE ORGONOMIC POTENTIAL BETWEEN CLOUD AND ITS ENVIRONMENT IS LOWERED."

Figure 5 represents the technique used by Reich to dissolve clouds.

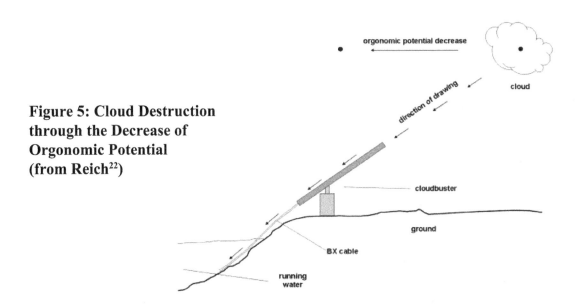

Figure 5: Cloud Destruction through the Decrease of Orgonomic Potential (from Reich[22])

Reich also explored how to dissolve fog that forms during autumn and winter. He developed a technique, somewhat different from the one used to create rain, which he called *foglifting*. He started to carry out the first experiments at the end of 1952 at Orgonon. He observed that an absorption of orgone in the atmosphere with the *cloudbuster,* applying a spiral technique from the zenith towards the horizon, gave the best results. This method disturbs the uniformity and the regularity of the fog bank. In fact, it is actually the lack of energetic differences in the orgone energy potential that favors the tenacity of the fog and its resistance to dissolving.

To disperse the fog, the regularity of its distribution must be interrupted by creating local energy potential differences. For the spiral *foglifting*, a very short absorption (a few seconds) in different points in the sky is performed. The absorption begins at the zenith, for 15-45 seconds. After this, the spiral movement proceeds slowly towards the horizon, with an intermittent movement, with brief stops of 5-10 seconds at each one. In most cases, to repeat this procedure 4 to 8 times is sufficient to dissolve a fog bank. After having carried out the spiral absorption, it is always necessary to stimulate an orgone flow form west to east that takes away the fog. This last operation should last from between half a minute to a maximum of 1 to 2 minutes.

Figure 6 represents the technique Reich developed for the spiral *foglifting*.

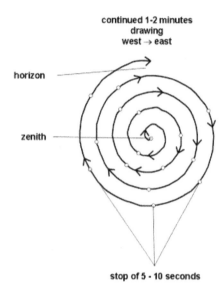

continued 1-2 minutes
drawing
west → east

horizon

zenith

stop of 5 - 10 seconds

Figure 6: The Spiral Draw Technique for Foglifting (from Reich[22])

Reich's Emphasis on Responsibility and Safety in Operations

Despite the scientific value and the usefulness of his discoveries, Reich immediately realized the potential dangerousness of the methods he had developed. One of the most important hazards is the long-distance effect of the operations. He observed strong reactions to *cloudbusting* activity he had performed at Rangeley in much distant areas, even up to 400-500 miles away. These effects are due to the continuity of the atmospheric orgone envelope.

Reich realized cosmic orgone engineering would have effects which could proceed beyond the borders of any individual state or nation, eventually making its rational social regulation an important international issue. In the atmosphere, where meteorological conditions form and change, there are no borders, no customs and documents to show. The modification of the weather conditions in a certain area therefore becomes a worldwide problem, with great and sometimes dramatic, often unknown repercussions for other regions. Reich warned that in order to avoid future chaos, it would be absolutely necessary to regulate all *cloudbusting* activities. Today, after fifty years of operations all over the world, outside of a few voluntary cooperative efforts, these measures have yet to take place.

During the operations, Reich always took care not to overdo the experiments, given that sometimes, he had observed the formation of small tornados and rapid wind changes. Once, following an operative error, he produced a prolonged and abundant rainfall.

He wrote an article in *CORE*[23] on the dangers and possible abuse of these methods, warning that only highly qualified people with a great sense of responsibility should be able to put the methods and protocols he had developed into practice:

> *"The day I had tried successfully to keep rain out from the Rangeley region, I spoke to our operator at Orgonon, Mr. Ross. He confessed to have been greatly impressed by the sense of power experienced in such operations as foglifting or rainstopping. I told him: The worst enemy of Oranur Weather control operations would be such feelings of power. The CORE operator, I explained again, is comparable to the steam engineer who directs the speed of the flow of steam with a single lever. He is, however, not the force itself that drives the train. He is only its trustee. And trustees must be fully conscious of their responsibilities, never forgetting what powerful force are entrusted to their care. To eliminate from the scene of CORE operations easy inflation of chronically deflated egos, seemed no less important than these operations themselves."*

No less important are the risks which the person carrying out the operation are exposed to. Whoever undertakes *cloudbusting* potentially absorbs excessive quantities of orgone energy into their body from the surrounding environment. The consequences can be headache, nausea, vertigo, even cardiovascular problems and sometimes even more serious conditions such as permanent paralysis, if the energy overload is very high. It is necessary to operate the *cloudbuster* with much caution and usually from a

**Photo 2 – Wilhelm Reich During a Cloudbusting Operation in the mid 1950s
(Courtesy of Wilhelm Reich Museum)**

distance, and only for short periods of time and for specific purposes. In addition, the excessive energy needs to be discharged from the organism of the operator.

Reich published in CORE[24] some simple rules to be followed during the operations in order to reduce the risks to a minimum, both for the operator and for the surrounding environment. In his own words, they are the following:

Rules to Follow in Cloud Engineering

1. Shed all ambition to impress anyone.

2. Never play around with rain making or cloudbusting. The OR envelope which you tackle while "drawing" energy from the atmosphere is an energetic continuum of high power. You may cause twisters. You may stir a forest fire into a wrong direction. You may do other damage without intending to do so. Never do anything unless you *must*.

3. If experimenting: it is important to observe and know why you are doing what, than to achieve immediate results. Slowly growing comprehension will secure later results safely.

4. In cloud engineering you do *not* "create rain" – you do *not* "destroy clouds" – briefly, you are not playing God. What you do is solely helping nature on its natural course.

5. Have your equipment, truck, etc. especially all metal parts well grounded into water, preferably flowing water. Lack of grounding imperils your organism.

6. Do not let workers draw OR energy *any* longer if they become blue or purple in their faces or feel dizzy. Exchange the workers; let them rest far enough away, and have their faces and arms always wetted down with fresh water.

7. Do not hold on to pipes or other metal parts while you draw OR. Always use a separate plastic or wooden handle to move equipment while drawing. Have your hands always well insulated with rubber or heavy cotton gloves.

8. Have signs put up in such a manner that no one is hurt by OR charges. Do not let people stand close by. Among them may be men or women who are ill and would run some danger to their health.

9. Never "drill a hole" into the sky right above you unless you aim for a long drawn rain.

10. When you feel a breeze or wind setting in due to your operation, stop drawing if the wind becomes too strong or even if it acquires the appearance of a twister.

11. If you wish to remove DOR clouds, *draw in the direction of run of the OR envelope.*

12. If you wish to DESTROY clouds or to stop rain, *aim at center of heaviest clouds.*

13. If you wish to make clouds grow heavier, draw from the vicinity of the smaller clouds, and leave the large or heavy clouds undisturbed.

14. If there are no clouds in the sky and clouds should be created, disturb the stillness or evenness of the OR envelope all around you by brief, sweeping draws and *draw mainly against the run of the OR envelope*. To create clouds you must create differences of OR energy potentials.[24]

A final point has been added in recent years to this discussion by Jamerling Ogg of PORE:

15. Under no conditions should you have any electrical wires, electrical instruments of any kind, or any radioactive material near a cloudbuster during any operation. Severe health problems may result.[25]

Chapter 4 Citations

1. The orgone energy does not have mass, and therefore cannot have either inertia or weight. Its concentration increases around planets and towards the Sun, which also excites the orgone into more active states. Matter derives from concentrated or frozen orgone. It is found everywhere in nature, though its concentration varies from place to place and depends on the meteorological conditions. The concentration of orgone can be influenced by the presence of clouds, humidity, temperature, and even by the time of day. It is attracted to, and can be bound to the atmospheric gasses and to water vapor, affecting their compositions and behavior. It is responsible both for the formation of various gaseous components of the air and of the movements of the air mass and humidity of the atmosphere. Orgone attracts itself and has a mutual attraction with water. Higher concentrations attract lower concentrations, until the maximum charge of the system has been reached, after which the orgone energy may dissipate spontaneously to lower levels, or even transform into electrical charge which is then discharged as a spark or lightning. Organic substances absorb this energy, while metals absorb it and then immediately reflect it. It spreads everywhere in rotating, undulating and pulsating waves. It is excitable by radioactive substances, which causes a reaction that has DOR as an end product. DOR is harmful to living beings and takes away humidity, oxygen and vital energy, causing a variety of illnesses. Orgone energy can be concentrated through an accumulator, constructed from alternating layers of organic and inorganic (metallic) materials. It is attracted and absorbed by the organic material, while the metallic layers promote its flow and build-up inside of the accumulator, which is lined by metallic walls. Concentrated orgone influences instruments like thermometers, electroscopes and the Geiger-Müller counter, and can be visually seen through special viewing devices, and even photographed.
2. DOR (acronym for Deadly Orgone) is the result of the degradation process of the orgone energy that has lost all its vital characteristics. If it is present in high concentrations, it can impart a sense of immobility and desolation on the area it hits, manifesting in a genuine lowering of the vital functions in the atmosphere, and everything in nature seems to assume a blackish or opaque aspect. The organism reacts to the DOR clouds with a rather serious state of illness principally due to an exhaustion of bioenergy. The Geiger-Müller counter can react in an unusual way to the presence of clouds that are highly DOR-charged. The reactions can vary: from a simple *jamming*, where the counter jumps to the top of the scale (100,000 CPM, or 20 MR/H on the portable counter SU-5 Tracerlab), to *failing* or *fading* where the counts fall rapidly to 30–40 CPM, settling down on 5-10 CPM. In the case of an atmosphere extremely charged with DOR, it can happen that the needle of the counter has an erratic behavior. It can at first show a value of 30-40 CPM, then jump, for example, to 500 CPM, fall to 70-100 CPM, and then jump to 10,000 or even 30,000 CPM.

3. Drought is the result of a prolonged shortage of rain. It is considered serious when the agricultural production in a given area is reduced by 10%, and it is catastrophic when the reduction reaches 30%. A prolonged drought can cause aridity. This is characterized by reduction in the frequency and quantity of rainfall to less than 250 mm per year (according to the Widtsoe classification). Desertification instead is the ecological consequence of drought and aridity that hits a particular area. It is characterized by scarce rainfall, by a continuous erosive process of the wind, and by the formation of sandy surfaces.

4. Reich, W.: *DOR Removal, Cloud-busting, Fog-lifting*, CORE, Vol. VI, No. 1-4, Orgone Institute Press, Rangeley, Maine (USA), July 1954.

5. Reich, W.: *Cosmic Superimposition*, The Wilhelm Reich Foundation, Rangeley, Maine (USA), 1951.

6. Reich, W.: *Expansion and Contraction in the Atmospheric OR Energy*, CORE, Vol. VI, No. 1-4, Orgone Institute Press, Rangeley, Maine (USA), July 1954.

7. Short for Orgone Operation.

8. Reich, W.: *OROP Rangeley, August 1-2, 1952*, CORE, Vol. VI, No. 1-4, Orgone Institute Press, Rangeley, Maine (USA), July 1954.

9. Reich, W.: *OROP Ellsworth, July 5-6, 1953*, CORE, Vol. VI, No. 1-4, Orgone Institute Press, Rangeley, Maine (USA), March 1955.

10. Reich, W.: *OROP Orgonon, July 23, 1953*, CORE, Vol. VI, No. 1-4, Orgone Institute Press, Rangeley, Maine (USA), July 1954.

11. Reich, W.: *OROP Boston, September 2, 1953*, CORE, Vol. VI, No. 1-4, Orgone Institute Press, Rangeley, Maine (USA), July 1954.

12. Reich, W.: *OROP Galactic Stream, Hancock, October 22, 1953*, CORE, Vol. VI, No. 1-4, Orgone Institute Press, Rangeley, Maine (USA), July 1954.

13. Reich, W.: *OROP Drought Atlantic Coast - Summer 1954*, CORE, Vol. VII, No. 1-2, Orgone Institute Press, Rangeley, Maine (USA), July 1954.

14. Reich, W.: *Expedition OROP Desert Ea. DOR Clouds Over the USA*, CORE, Vol. VII, No. 1-2, Orgone Institute Press, Rangeley, Maine (USA), March 1955.

15. Reich, W.: *Report of OROP Desert Ea Survey of Tucson Area. September 18, 1955*, CORE, Vol. VII, No. 3-4, Orgone Institute Press, Rangeley, Maine (USA), December 1955.

16. Reich, W.: *Contact with Space*, Core Pilot Press, Rangeley, Maine (USA), 1957.

17. Reich, W.: *OROP Hurricane Edna*, CORE, Vol. VII, No. 1-2, Orgone Institute Press, Rangeley, Maine (USA), March 1955.

18. Reich, W.: *OROP Children's Parade, August 8-9, 1953*, CORE, Vol. VI, No. 1-4, Orgone Institute Press, Rangeley, Maine (USA), July 1954.

19. Marcovicz, D.M.: *Über Wilhelm Reich. Viva Kleiner Mann*, Nexus Verlag, Frankfurt, 1987.

20. Eden, J.: *CORE Manual. Cosmic Orgone Engineering*, Careywood, Idaho (USA), 1986.

21. Reich, W.: *DOR Removal, Cloud-busting, Fog-lifting*, CORE, Vol. VI, No. 1-4, Orgone Institute Press, Rangeley, Maine (USA), July 1954.

22. Reich, W., *ibid,* 1954.

23. Reich, W.: *OROP Boston, September 2, 1953*, CORE, Vol. VI, No. 1-4, Orgone Institute Press, Rangeley, Maine (USA), July 1954.

24. Reich, W.: *Rules to Follow in Cloud Engineering*, CORE, Vol. VI, No. 1-4, Appendix A, Orgone Institute Press, Rangeley, Maine (USA), July 1954.

25. Acronym for *Public Orgonomic Research Exchange*, founded by Jamerling Ogg in 1995. Located in Boothbay, Maine, it is primarily devoted to the spreading and exchanging of information on orgonomy and on Reich's research.

Cloudbusting After Reich

*"Desert souls will enhance desert development; and
desert development will increase DOR or staleness in the
human emotions."*

Wilhelm Reich

After Reich's death, there was an almost complete standstill in the field of orgonomic research which lasted for about ten years, during which orgone physics and especially *cloudbusting* seemed to have been forgotten. The subject remained of material interest only to those passionate few who carried out personal research in a private manner.

Only since the middle of the 1960s did the research and experiments begin once more on a larger scale, at first only isolated, but then becoming more and more widespread and frequent. Among those who have distinguished themselves more than any others in studying, implementing and spreading of orgonomy in a pioneering manner, the following persons stand out: Richard Blasband, Courtney Baker, and John Schleining, who were former members of the *American College of Orgonomy*[1], in addition to independent researchers such as James DeMeo, Robert Morris and Jerome Eden.

Over a period of more than 30 years, they carried out a large number of experiments, both individually and cooperatively, mostly in the United States. During these years, many of the results were published in the *Journal of Orgonomy,* the periodical edited by the *American College of Orgonomy*. They principally concern field experiments aimed at eliminating the presence of DOR in the atmosphere; reducing drought, heat and smog; producing rain and snow; and even one case of diverting hurricanes. Some of these publications, in addition to providing detailed information on the operations, also report the statistical results, which show a very significant correlation between *cloudbusting* and the changes of the atmospheric conditions.

Of particular interest are the experiments conducted by Charles R. Kelley[2] in the United States towards the end of the 1950s. He carried out a series of interventions to create rain in the region of Westport and Stamford (Connecticut) between October 1958 and August 1960. He documented all his operations with photographs that clearly show the changes in the size of individual clouds and cloud systems under the influence of the *cloudbuster*. It rained in all cases within 36 hours from the start of the intervention, carried out on days when the weather forecast had not predicted any rain for several days. He calculated the probability for rainfall by chance to be one in a thousand. He published the results of his experiments in a report[3] with many of the photographs related to the development of cloud systems and the procedures used. Photo 3 shows a before and after picture of an experimental cloud dissipation operation performed with the apparatus on top the Atlantic Power Building in Stamford in the summer of 1960. The "X" indicates the target of the *cloudbuster* at the outset. The lower picture shows the same scene 10 minutes later. Control clouds are on either side.

During this period, there were also experiments conducted in Italy by Bruno Bizzi[4] and his group, and by Walter Hoppe in Israel[5]. However, the results of these experiments were never published.

**Photo 3 – A Before (Top) and After (Bottom) Picture of an Experimental Cloud Dissipation
Operation Performed by Charles R. Kelley, Stamford (Connecticut), Summer 1960
(Courtesy of Charles R. Kelley)**

Richard Blasband's Research Activities

Starting in 1965, Richard Blasband and his colleagues carried out numerous field experiments, primarily aimed at verifying the effectiveness of *cloudbusting* in producing rain during times of drought, in diverting of hurricanes, the production of snow in tourist locations at high altitudes, and in the extinguishing of forest fires.

Parallel to the experiments, one of the most important things that needed to be developed was the establishment of scientific methods for evaluation of the results, in order to verify if *cloudbusting* was indeed capable of altering atmospheric conditions, or if the occurring changes were only due to natural and spontaneous variations in the atmosphere.

One of the biggest problems facing scientific research on methods of climate intervention is that of establishing adequate control procedures. In general, this is a typical problem of all research fields, but in meteorology it is especially difficult to create laboratory conditions that could perfectly simulate the conditions existing on a large scale. Therefore, evaluation of measured weather data almost always remains the only way to possibly demonstrate the effectiveness of the operation. However, to evaluate and determine the effectiveness, it has traditionally been necessary to carry out many operations over a certain number of years, for prolonged periods and always at the same place.

Thanks to satellite images, it has been possible within the last twenty years to follow and evaluate individual operations in a more precise and reliable manner, while it was previously very difficult to know the actual extent and nature of the atmospheric systems one was dealing with. New information is also drawn from detailed weather maps which the meteorological services supply on the trend of atmospheric frontal systems being evaluated. Through such use of satellite images and weather maps, it has become easier to immediately evaluate the atmospheric energy state and the effectiveness of an operation in real time, even at a distance of thousands of miles. Besides, those pictures from space can today be obtained in simple ways from different sources: by computer, by fax, and by cable or satellite television. The images are generally updated every half hour, making it possible to follow the changes in cloud formation or development of storm systems in an almost immediate way.

Blasband[6] (shown in Photo 4 with his cloudbuster *Silverhorse*) proposed a methodology for verifying if the *cloudbuster,* when pointed at the sky, can really influence the atmosphere both in the removal of DOR, in the dissolving of clouds, and in the production of rain.

He developed a very simple protocol to evaluate the effectiveness of the removal of atmospheric DOR. The test must be carried out on a day when there is a moderate or high concentration of DOR[7] and little air movement. Some minutes after the start of the *cloudbuster* absorption, a fresh breeze should come up. Sometimes, the flow of cold winds can last for some hours after ending the absorption. At the same time, an increase in the luminosity of the sky can often be measured using a light meter. The difference between the initial and the final readings of this parameter is a general indication of the effectiveness of the operation in removing the DOR.

In case the dissolving of a cloud is studied, it is necessary to select a day with a calm sky and the presence of some cumulus clouds. These clouds must have a distance between them of at least the same length as that of the clouds themselves. After pointing the *cloudbuster* at the pre-selected cloud, which is generally chosen by chance by an impartial observer, the dissolving should begin within a few minutes after the start of the absorption. The thinning process is irreversible, and the cloud should disappear completely within 5-20 minutes. The effect is obvious and unmistakable to the naked eye.

Photo 4 – Richard Blasband With the Cloudbuster *Silver Horse* in Bucks County (Ottsville, Pennsylvania), at the Start of the 1980s (Courtesy of Richard Blasband)

Finally, an operation carried out to generate a storm system should bring about rain within 8 to 48 hours. The total results can vary from a local storm to a heavy rainfall over a vast area, even up to thousands of square kilometers. These effects depend on the natural weather which is already present, on the power of the apparatus used, and on the duration of the operation. Contrary to chemical-physical phenomena, where there is generally only a short delay between the cause and the effect, atmospheric orgone energy processes exhibit a delay in the response time when being externally excited, and therefore in the appearance of the effects. Therefore, the delay between the start of the operation and the creation of clouds and the production of rain can vary from several hours to a couple of days, and in some cases of severe drought even any number of days.

To evaluate the results in this case, since the response time of any effect is quite long, one method is to rely upon the official weather forecast. These forecasts are generally accurate on the short-term. For a period of 24 hours, the accuracy is between 80 and 85%, while for a period of 48 hours, it varies between 70 and 75%.

Blasband established the following criteria to evaluate the effects of an operation to produce rain in a reliable way:

♦ It is necessary to select a day where the weather forecast for a given area has predicted rainfall to occur with a probability of less than 10% within the next 24 hours, and no precipitation for 48 hours.

♦ In order for the *cloudbusting* to be deemed effective, there must be rain, even very light, within 48 hours from the onset of the operation.

However it is very difficult to evaluate the results of an operation intended to intensify a storm already predicted by the weather forecast. This is because the weather forecast rarely supplies information on the quantity of rain that should fall in a determined area. Therefore it is not possible to carry out comparative and quantitative evaluations, unless the forecast predicts a light rain and the operation produces a heavier storm.

The Removal of DOR and the Reduction of Drought

One of the main areas of intervention to which Blasband and his colleagues always dedicated a large amount of their efforts were the removal of DOR from the atmosphere, the reduction of drought, and the decrease of high temperatures that have hit certain parts of the United States in the last thirty years.

DOR is possibly the most widespread disturbance that can influence the natural behavior of the atmosphere. Reich observed that DOR was produced through the reaction of orgone energy with radioactivity, with X-rays, and with electromagnetic instruments. The result is firstly a hyper-excited state of orgone called the *Oranur reaction*. If the causative environmental conditions don't change, the hyper-excited state of the orgone transforms into a secondary stagnant and dulled condition, a state without life, hungry for oxygen, water and orgone, now called *DOR*. The overall process results in a general loss of water and oxygen in the atmosphere.

Drought, extremely high temperatures, and an increase in desertification are some of the more obvious climatic changes that have been recorded in the last ten years on a global level. Towards the end of the 1970s, approximately 43% of the Earth's surface was either a desert or a partial desert[8]. In 1975 it was estimated that the fertile and productive land would diminish at a rate of 14 million acres per year[9] (around 5.6 million hectares). Towards the end of the 1980s, the annual increase in desertification had reached about 7 million hectares[10].

Trenberth of the *National Center for Atmospheric Research* in Boulder, Colorado, estimates that the proportion of land classified as experiencing very dry conditions in 2002 was around 30%, involving areas of Europe, Asia, Canada, and parts of Africa and Australia[11].

A study by the *National Academy of Sciences*[12] highlighted how much the irregular warming of the atmosphere had increased in the last century. From the available data, it appears the Earth is experiencing the hottest century of the last thousand years, and one of the hottest of the last 10,000 years.

Jenkins of the *Hadley Centre for Climate Prediction and Research Meteorological Office* in the UK considered the variation of the temperature as one of the most important parameters for evaluating the climatic changes on a global level. Figure 7[13] shows the annual average near-surface temperature variation trend of the planet from 1860 (year in which the first global measurements were taken) until 1996. The unbroken line represents the 10-year running mean.

As you can see in the Figure 7 graph, 9 of the 10 warmest years since 1860 are found in the years between 1980 and 1990, while the hottest year is 1995. In addition, starting from 1860, the average temperature of the earth has increased by 0.6 °C, with a rise in the sea level of 10/25 cm in the last 100 years.

A reduction in rainfall has also been recorded in the tropical and subtropical areas coincidental to these temperature changes, with an increase in the frequency of extreme climatic events such as floods, storms, and episodes of excessive hot or cold temperatures.

It is also evident that these changes have been accompanied by an increase in the instability of the Earth's crust. The graph in Figure 7 shows a sudden increase in the global temperature which started in the middle of the 1970s. Parallel to that, a strong planetary seismic activity was registered during those years. In the first half of 1976, there were 10 large earthquakes[14], double the annual average. Three earthquakes, which had a magnitude of more than 8 on the Richter scale, were recorded in the same period. This number is quite high when compared to the annual average of only one such event. Blasband pointed to the intensified activity of the Earth's crust as a possible cause for concern, since the increase in the number of earthquakes could indicate that even the crustal membrane covering the planet could be influenced by an *Oranur* reaction in the same manner as the atmosphere.

Yet the reasons for these great changes are still largely unknown. They could be of terrestrial as well as cosmic origin. Many believe that it is a combination of the two. So far as terrestrial causes are concerned, the relationship between this rise in atmospheric temperatures and the rapid increase in human population, deforestation, agriculture, and industrial production and pollution since the start of

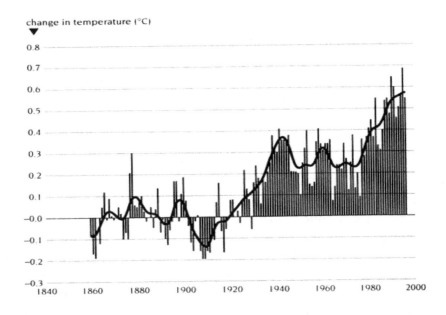

Figure 7: Global Surface Temperatures 1860-1996 (from Jenkins[13])

the last century is evident. In addition, there is proof of the irritating effects of nuclear radiation and other toxic pollutants on the atmospheric orgone envelope. During *cloudbusting* operations carried out close to nuclear power plants, Blasband et al.[15] experimentally confirmed that these disturb the atmospheric orgone continuum, affecting both the formation and the development of cloud structures.

Other possible causes for the formation of DOR in the atmosphere are natural cosmic cycles related to the influence of the Moon, the Sun or galactic sources. Even though very little is known of the latter causes, and the results are still not fully confirmed, there is proof that sunspots and the Moon have effects of excitation and expansion on the orgone energy continuum, and consequently on the Earth's climate. Blasband[16] even suggested the probability that the Earth is crossing an area of space with a higher density of DOR, which could be one of the major reasons for the desertification process currently underway on our planet.

It is therefore difficult to evaluate the actual extent of the various causations for the formation of DOR in the atmosphere and the possible long-term remedies. In case the climatic problems of the Earth's atmosphere are of cosmic origin, very little can be done.

Meanwhile, the solution to the problem of DOR created by human beings is already being implemented, even if it is to a very limited extent. The most important changes in this sense concern both today's technology that pollutes and irritates the atmosphere in a continuous and senseless way, and the formation of people's armored character structures. Reich noted a strong similarity between desert plants and armored people. Both were prickly on the surface, and emotionally empty and desert-like inside. In this case, the solution lies principally in changing the character structure of humanity, by the elimination of emotional armoring. Reich was very clear on this point, believing in fact that human emotional armor, through its actions and inventions, creates a very similar desert environment.

From the orgonomic point of view, the primary cause of drought and desertification is the formation and accumulation of DOR in the atmosphere. Drought and desert areas are characterized by an atmosphere that is heavy with DOR, and in a chronic state of expansion. Their atmospheres completely lack the contraction and discharge functions and the cyclonic movements, which are fundamental for the condensation process (humidity and rain).

In order to better explain the phenomenon of DOR formation in the atmosphere, Blasband[17] made a comparison with the research carried out in the field of biopsychiatry. Based on Reich's discoveries, he noted that the emotional armor of human beings is the result of a peripheral muscular contraction, to fend off an internal emotional charge generated within the core of the organism, deep in the solar plexus region. The emotional armor holds and blocks the energy flow out into the organism. The immobilization is eliminated through the breaking and dissolving of the emotional-muscular armor, allowing the energy to once again move freely. DOR which may be present in large concentrations within the organism can be partly absorbed out by a *DOR-buster*, which is essentially a small *cloudbuster*. Correspondingly, the presence of DOR in the atmosphere is an indication for a state of atmospheric armor which blocks the atmospheric orgone energy flow and pulsation. Absorbing the DOR with the *cloudbuster* is basically an attempt to soften the block and dissolve the armoring in the atmosphere, in order restore the natural movement of the energy.

Blasband noted that to break down the atmospheric armor, it was necessary to absorb the DOR while trying to keep the atmospheric orgone continually moving, until the atmospheric armor had been dissolved — but also not to over-work the *cloudbuster*, which can result in the unwanted return of chronic expansion and stasis. This point is of fundamental importance, and is often the reason for the failure of many *cloudbusting* operations. One must not end the absorption too soon, nor work too long, as otherwise the DOR returns to its original level and reforms the blocking armor in the atmosphere. In some cases of hard desert or prolonged drought, it may be necessary to operate for many days and in different places in order to break down and dissolve a widespread and chronic DOR condition. In other cases, only a few minutes of absorption will trigger a widespread dissolving of atmospheric DOR and armoring, with a return of natural rains.

In areas where large quantities of DOR are present, great variations in the atmosphere's physical parameters can also be observed. It is possible to observe very low monthly mean humidity values, where condensation can occur only at altitudes much higher than under normal conditions. In such cases the evaporation rate of the Earth's surface is higher, and it is possible to measure extreme readings of high and low temperatures and high pressure, even for long periods.

Blasband[18] saw that, in all such cases where DOR was present, the energetic charge and discharge of the atmosphere occurred only outside the drought and desert area, due to a diversion of energy streams. For example, it was established[19] that under normal atmospheric conditions, the behavior of the climate in the Pacific Northwest of the United States was determined by the superimposition of the orgone energy streams that flow both from the southwest and the northwest. In the case of a prolonged deviation of these main orgone streams, generally drought and the development of desert will occur in the affected areas.

As mentioned above, DOR acts harmfully on the atmosphere, absorbing the humidity in the air and interfering with the cohesive and attractive forces of the orgone, which work at the basis of cloud formation. In turn, this mechanism sets the process into motion that gives rise to drought. All these phenomena are functions of the quantity of DOR trapped in the atmosphere.

Therefore, even if the direct removal of DOR with the *cloudbuster* provides a temporary relief from its toxic influence, it is necessary to emphasize that in order to completely resolve the problem and to restore the natural environmental conditions in a certain area – or even globally – it is of fundamental importance to understand and resolve the original and primary causes of the DOR.

The operations, generally carried out to produce rain, can bring water and humidity to an affected area, but do not necessarily resolve the problem of drought, which tends to return after some period of time. Blasband has pointed out that the primary reasons for the formation of DOR (human activity and cosmic influences), for drought and for the formation of deserts will not be addressed by the *cloudbusting* operations alone, other than in a very marginal way.

It is obvious, considering these global forces that influence the planet, that the intervention procedures on the weather performed with the *cloudbuster* have an effect that is essentially local and above all, temporary or symptomatic.

Among the most significant operations by Blasband are those carried out towards the end of the 1960s, when satellite images were used for the first time to follow and document the development of cloud systems in real time.

Operation *Heatwave*[20] was intended to reduce the high temperature and humidity levels that had been recorded in the area of the *Middle Atlantic* states during the month of July 1977. On the 16th and 17th of July, Blasband and his group undertook some brief absorptions with the *cloudbuster* with the aim to eliminate the DOR present in the atmosphere.

However, the absorptions did not have the desired effect. On July 19th, the tubes of the *cloudbuster* were extended to their maximum length and grounded in a lake situated close to the New York area. The goal was to absorb in a stronger manner in order to create a contraction in the atmosphere that was chronically expanded, and to consequently bring about rain. The weather forecast of that day estimated the probability of rainfall at between 0% and 20% for the subsequent days. Despite this prediction, strong storms and abundant rain occurred in the area during the night. On July 21st, contrary to the forecast, the temperature lowered by about 10° F[21], and there was abundant rainfall in all the *Middle* and *Southern Atlantic* states.

The ability of the *cloudbuster* to initiate an orgone energy flow where there had been almost no movement before, was evident from the satellite images. The map in Figure 8 shows the cloud cover present in the northeastern area of the USA on the 19th of July at 11.01 am, prior to the start of the prolonged absorption. The previous maps did not show significant changes, and the weather forecast did not forecast any short-term changes, confirming the chronic stasis situation in the atmosphere in that part of the USA.

Figure 9 shows the situation two and a half hours after the end of the operation on July 19th. There is an obvious change in the cloud system, with a break-up of the wide mass of clouds that hung over western Pennsylvania. The clouds started to move eastward and showed a tendency to build up and reform to the southwest and northeast.

The map in Figure 10, taken an hour and a half after Figure 9, shows a rapid development of all systems.

Finally, Figure 11 presents the situation at 10.01 am on the 20th of July. This map shows a strong concentration of all the clouds into a large mass centered over the absorption point, determined by the location of the *cloudbuster*, and responsible for the abundant rainfall on the 21st of July.

Therefore, the effects of the *cloudbuster* in this operation are evident and well-documented by the satellite images. After this positive result, the operations were suspended. This was an error, as the drought returned in September with an unforeseeable strength in many parts of the country.

A second significant example for the influence of *cloudbusting* on the weather and on storm processes was carried out at the beginning of 1981[22].

In the months between October 1980 and the end of January 1981, the entire United States were subjected to drought. The responsible atmospheric systems were considered the most unusual ever

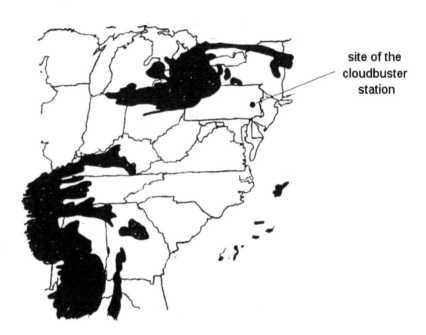

site of the cloudbuster station

19 July 1977 11.01 am

Figure 8: Operation Heatwave, Cloud Cover before Beginning the Draw (from Blasband[20])

site of the cloudbuster station

19 July 1977 02.31 pm

Figure 9: Operation Heatwave, Cloud Cover 2.5 Hours after Ending the Draw (from Blasband[20])

registered. Two vast and persistent high-pressure systems were situated over the *West Coast* and the Rocky Mountains in a particular way, blocking the movement of the atmospheric systems that would bring rain, and extending over the central part of the continent, towards the *Mid West* and the *East Coast* states. The polar jet stream had hence shifted towards central Canada, taking with it the low-pressure systems which were responsible for much of the precipitation in that area.

Blasband and his group decided to intervene with operations aimed at creating rain specifically in the central eastern part of the United States. The operation started on January 28th and continued daily until the 31st, resumed between the 7th and the 9th of February, and then continued from the 11th to the 19th of February.

During this period, there was much rain despite the fact that the weather forecast had reported only scarce and isolated showers. From the maps[23] showing the monthly trend of the level of drought, one could see that after a while, the humidity values went back to normal.

In addition, a marked lessening of moderate drought conditions in some areas, and a complete disappearance of the extreme drought conditions in others was notable in January. The positive results obtained caused a suspension of the operations. This once again was an error, because the drought restarted in March. The operations were resumed on March 24th, with 3 to 4 absorptions per week until mid April. Table 1 shows the number of operations carried out in the months of February, March and April, the number of rainy days, and the effect on the drought.

It is obviously not possible from this data to undertake a statistical analysis

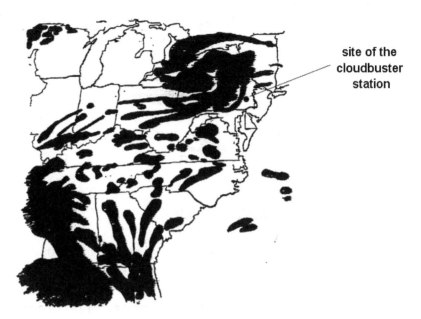

site of the cloudbuster station

19 July 1977 04.01 pm

Figure 10: Operation Heatwave, Cloud Cover 4 Hours after the Ending the Draw (from Blasband[20])

20 July 1977 10.01 am

Figure 11: Operation Heatwave, Cloud Cover 22 Hours after the Ending the Draw (from Blasband[20])

and draw quantitative conclusions. However, the close relationship that exists between the number of operations, rainy days and the effect on the drought is notable. In addition, very often the weather forecast had predicted a light rain, while actually a strong and abundant rainfall occurred.

Month	Cloudbusting days	Rainy days	Effect on Drought
February	15	6	reduction
March	3	1	increase
April	10	5	reduction

Table 1

Another example where the positive results of the operations are evident, is the intervention carried out in the spring of 1985, in the area surrounding New York[24].

In that period, the water reserves of New Jersey and particularly New York had fallen to 61% of the normal level. This situation was the result of months of scarce rain. By the middle of May, the situation was serious enough to declare a state of emergency, and to initiate water rationing in the city of New York and a large part of New Jersey. The atmosphere was oppressive, with an elevated DOR index, and a large quantity of *Oranur* and DOR blocking a chronically overcharged and expanded atmosphere. Most of the humidity coming with storm systems from the west, was completely absorbed by the DOR and dissipated within the atmosphere.

Blasband decided to start operations on April 18th to try to bring rain to that area. The *cloudbuster* was set up to absorb and remove the DOR and stimulate the local energetic contraction of the atmosphere. The weather forecast for the area from Philadelphia to New York had not predicted any rain for at least 3 or 4 days. The next day, there were rains and storms over the whole region. Based on the experience acquired during previous years, Blasband decided to stop the interventions when the expected results were obtained, and to resume the absorptions when the drought tended to reappear once again. In this way, the *cloudbusting,* carried out at pre-established intervals and for a longer period of time, allowed a complete elimination of the drought that had hit the area within 3 months. Six weeks after the start of the operations, an inversion of the climatic characteristics could be noted. The more or less chronic drought situation changed to atmospheric conditions more favorable to rain, and there was abundant rainfall before mid-May. The operations often consisted of a combination of interventions addressing both the contraction in the atmosphere and the creation of rain. In the latter case, the *cloudbuster* was set up in such a way so as to generate storm systems in the west, thus creating a flow from west to east which brought rain to the area where the *cloudbuster* was stationed. All operations were constantly monitored with satellite images and the drought index maps. For the period in question, they show a net change in the drought index for the state of New Jersey, going from beyond "extreme" in April to "normal" in June. Figure 12 shows the quantity of rain that fell during each week in April, May and June, in the area surrounding the location of the *cloudbuster*.

As can be seen in Figure 12, the quantity of rain that fell in the region around the *cloudbuster* clearly shows a relationship between the operations and the changes in atmospheric conditions. From this data and the drought index maps[25], it is without a doubt that *cloudbusting* effectively produced a change in the weather.

One last example refers to the experience of Schleining[26] who carried out prolonged interventions in Southern Oregon with the goal to reduce the strength and the persistence of the drought that had affected the region since 1984.

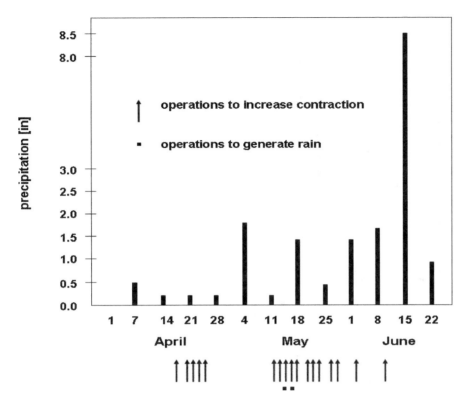

Figure 12: Storm Enhancement Procedures (from Blasband[24])

The quantity of rainfall until 1991 was greatly below normal, with a deficit of around 46.2 inches. In 1992, the quantity of rain that fell until June reached about 47% of the normal value, with a deficit of around 10 inches. The rivers and water wells were drying up, and the farmers feared great damages to the crops due to the lack of rain. In addition, record temperatures higher than 110°F were registered for that area, with an increased risk of forest fires.

Schleining decided to intervene with an operation aimed at reducing both the DOR present in the atmosphere and the drought. It was carried out in a way to bring the main atmospheric flows and jet streams into the area, which had shifted from its usual trajectory because of a chronically overcharged and expanded atmosphere.

He worked continuously for 17 days, from June 24th to July 9th. The atmosphere reacted positively to the interventions, so much so that the DOR index lowered significantly and the main atmospheric flow responsible for bringing rain to the area, moved directly overhead of the *cloudbuster*, restoring the natural flow conditions. During the eight days of absorption, 2.4 inches of rain fell, which was considered a surprising result under those conditions. Figure 13 shows the trend of the daily rainfall and the reduction in the average daily temperature from the norm for the months of June and July.

Also in this case, the graph shows a close relationship between the operation of the *cloudbuster* and the development in rainfall quantity.

The Deflection of Hurricanes and Other Interventions

Among the various interventions that Blasband and his colleagues carried out were those aimed at diverting a hurricane, putting out forest fires and generating snow at high altitudes. These operations are much more complex and difficult to undertake.

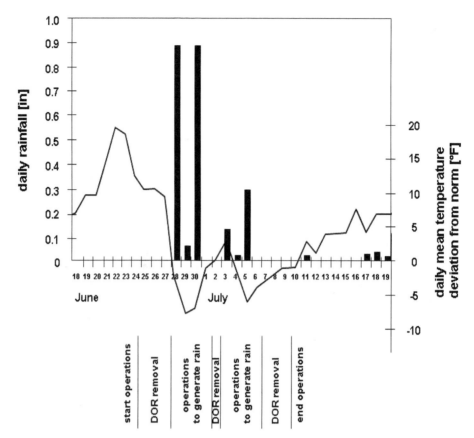

**Figure 13: Daily Rainfall and Daily Mean Temperature
Deviation from Norm (from Schleining[26])**

Blasband[27] warned of the complexity and dangerousness of these interventions, because if the operations were performed incorrectly or if some parameters unexpectedly went out of the control of the operator, there could be potentially disastrous results, with the loss of human life and social destruction. These types of natural events needed to be approached with a great deal of caution and an enormous sense of responsibility, especially when working to alter the course of a hurricane.

From the orgonomic point of view, hurricanes are formed by the superimposition of two streams of orgone energy which are initially moving in different directions but then form vortices of spiralling energy with an anti-clockwise rotation in the Northern Hemisphere and a clockwise rotation in the Southern Hemisphere. Reich noted that the energy currents responsible for this phenomenon were the Earth's Equatorial Orgone Stream, which flows from west to east, and the Galactic Orgone Stream, which forms a 62° angle with the Earth's equator.

Hurricanes carry in their paths from the sea towards dry land great quantities of water associated with an enormous quantity of energy. Blasband suggested that hurricanes probably have the task of discharging the high orgone potential that forms in the tropical atmosphere during summer, and especially to eliminate DOR from specific areas.

The only attempt by Blasband to divert the course of a hurricane was made in 1967. He tried to change the path of hurricane *Doria* which was threatening disaster to the *Middle Atlantic* states. The only previous attempt was undertaken by Reich[28] in September 1954, when he tried to deflect the course of hurricane *Edna* that was threatening the *Middle Atlantic* states and *New England*.

The operation Reich carried out was based on fundamental orgonomic principles. First of all, he tried to increase the flow of atmospheric orgone energy from west to east, to push the hurricane towards the sea. Also, he withdrew energy from the hurricane core by pointing the *cloudbuster* directly at its center, in order to weaken and disturb its inner movement.

The weather forecast had predicted that hurricane *Doria* would hit the states of Virginia, Maryland, Delaware and New Jersey at around 8 am on the 16th of September with a speed of about 90mph[29]. Blasband decided to intervene at 11 am on the 15th of September, pointing the tubes of the *cloudbuster*, completely extended and deeply immersed in water, directly at the center of the hurricane that was situated to the southeast of the absorption site at that time. The goal was essentially to weaken it by absorbing part of its energy. After that, the tubes were moved in a manner to create an energy flow in the atmosphere from the northwest that could divert the course of the hurricane by taking it far away from the coast out to the sea. This part of the operation was able to move the hurricane 60 miles off the Atlantic coast. Afterward, Blasband once again pointed the tubes of the *cloudbuster* at the core of the hurricane, in order to further weaken it and to eliminate it completely. Following this last operation, the hurricane started to move in a strange and irregular way, turning back towards the coast. At 2 am on the morning of September 17th, the radio weather forecast declared that *Doria* was no longer considered a hurricane and had definitively started to move away from the Atlantic coast at a speed of 45mph. Figure 14 shows the hurricane's path during the *cloudbusting* operation.

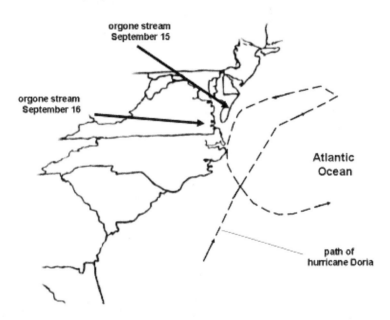

Figure 14: Path of Hurricane Doria, September 1967 (from Blasband[27])

In considering more complex and difficult activities, there is the ending of forest fires caused by drought and high temperatures, as characterized by the work of other groups of researchers. Among the most important operations carried out, we can find that of Eden in 1979[30] and Schleining in 1987[31]. The operation by Eden and his group was intended to reduce the drought and put out great fires that had developed in the northwestern part of the United States in August 1979 (Photo 5). Because of the drought, the water level of rivers and streams was rapidly falling, while the fires were spreading everywhere, destroying thousands of acres of woodland and some inhabited areas.

The operation started on the 4th of August 1979 and continued for 9 days. They absorbed with the goal of bringing humidity from the Pacific Ocean inland. Eden tried to create a difference in the orgonomic potential to allow clouds to form and grow over the area of the fire. In addition, he tried with targeted absorptions to attract the Galactic Orgone Stream to the region, as to increase the possibility

Photo 5 – Jerome Eden With Cloudbuster *Bluebird* During *Operation Firebreak*, August 1979.
Reproduced in Color on Rear Cover.
(Photo Taken by Desiree Eden, Courtesy of Jonathan Coe)

of rain in the area surrounding the *cloudbuster*. On August 10th the first rain started to fall, while the previous day's weather forecast had estimated the chance for precipitation at about 0 to 10%. In total, more than an inch of rain fell which considerably reduced the risk of fire in that period.

The operation undertaken by Schleining in the summer of 1987 was aimed at reducing and putting out the great fires that had developed in the State of Oregon. On August 31st of that year, a great lightning storm produced more than 10,000 lightning bolts, which started around 2,000 fires in the surrounding forests. The fires spread very quickly, thanks to the drought and the dryness to which the entire area had been subjected. It was estimated that the cost of putting out the fires was about $5 million a day, with the mobilization of more than 5,000 fire fighters. Within a few days, 20 counties of California and 10 of Oregon had declared a state of emergency. Schleining's plan of intervention included operations to increase the humidity in the area, lower the high temperatures, to reduce the winds and also increase the orgone potential to the west, which would create clouds and bring rain to the areas where the fires were blazing. The operation consisted of three interventions spread out over the months of September, October and November. During all these interventions, both the wind velocity and the temperatures were reduced, and the humidity in the air was increased. However, there was only a small quantity of rain. It was believed, the presence of nuclear waste storage facilities, atomic reactors and factories for the production of nuclear weapons in the nearby Hanford area of Washington State acted as a blocking mechanism against moisture coming into the region. Similar observations have already been made by other operators, that nuclear facilities disturb the surrounding atmospheric orgone continuum and can negatively influence the development of cloud structures.

Figures 15 and 16 show the surface area in acres that was lost per day to the fires during the first (Figure 15) and the second (Figure 16) intervention.

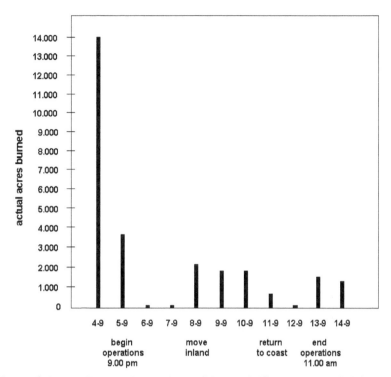

Figure 15: Number of Acres Burned per Day, Phase 1 (September 4-14, 1987) (from Schleining[31])

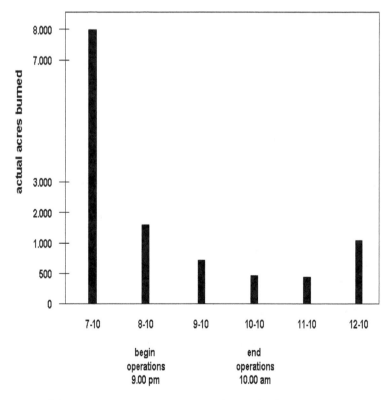

Figure 16: Number of Acres Burned per Day, Phase 2 (October 7-12, 1987) (from Schleining[31])

In Figure 15, the strong decrease in the surface area destroyed by the fires during the first phase of the operation is evident. It went from 14,000 acres per day to zero within two days from the start of the intervention. Figure 16 shows the decrease of acres burned per day during the second phase, from 8,000 acres on the day before the intervention, to 620 acres two days after the start of the operations.

At the end of the operation, the weather forecast of the *National Forest Service* on November 24th reported that all the forest fires had been contained and put out.

Another example for the application of *cloudbusting* is the activity of creating snow in the mountains. The only specific intervention for snow-production documented in the literature is the one carried out by Eden[32] in December 1978 in the ski resort of Schweitzer, situated on Mount Selkirk, Northern Idaho.

Eden intervened at the request of the company that managed the ski lodge and facility, to bring snow to the area. This operation was one of the first to be carried out in a mountainous area at low temperatures. In this case, there were problems never faced before, such as working with the grounding in snow instead of running water. Holes were dug in which to immerse the tubes of the *cloudbuster,* which were then covered with snow.

In the previous week, fine dry hail had fallen in the area with very little snow. The presence of DOR was very high, and that was probably responsible for the lack of snow. To make things worse, the southwestern front had bypassed the area, moving both to the north and the south. On the 16th of December, Eden started the operations with minimal absorptions, and at the end of the first day, 2.5 inches of snow had fallen on the basin. He continued the absorptions to remove DOR from the atmosphere. On the 18th of December, another 1.5 inches of snow fell. An abundant nightly snowfall of 9 inches was also registered in nearby areas. The operation continued until the 23rd of December. At the end of the operations, between 6 and 9 inches of snow had fallen on the mountain ski region.

This operation demonstrates that the *cloudbuster* can work well even at high altitudes and under extreme conditions without the presence of running water.

Chapter 5 Citations

1. The *American College of Orgonomy* has its head office in Princeton, New Jersey. It was founded in 1968 by Elsworth Baker, student and follower of Reich, with the goal of promoting and spreading research activities in orgonomy. It produces the *Journal of Orgonomy* every six months which publishes the results of research conducted by the Institute as well as by independent researchers.
2. Charles R. Kelley, student of Reich, founded the *Radix Institute* at the start of the 1960s. The activities of the Institute are principally based on the development of individual and group techniques derived from Reich's original discoveries on the reaction between emotions and the muscular armor. He published *The Creative Process,* a journal that for five consecutive years after Reich's death, starting from June 1961, was the only scientific journal publishing results of research carried out in the field of orgonomy.
3. Kelley, C.R.: *A New Method of Weather Control*, Kelley/Radix Publications, Vancouver (USA), 1961. Also see Kelley, C.R.: *Eine Neue Methode der Wetterkontrolle*, Plejaden, Berlin (Germany), 1985. Also reprinted in Kelley, C.R.: *Life Force: the Creative Process in Man and in Nature*, Trafford Publishers, Victoria, BC (Canada), 2004.
4. Bruno Bizzi worked as a psychiatrist at the neuro-psychiatric clinic of the University of Bologna. Later he acted as vice director of the Lolli Hospital at Imola. He was the first, together with Prof. Chiurco, director of the *International Research Center on Pre-Cancer Conditions* at the University of Rome, to introduce the use of the orgone energy accumulator for the cure of cancer to Italy.
5. Walter Hoppe, psychiatrist, student and co-worker of Reich, was one of the first to experimentally verify the effectiveness of the orgone accumulator for the cure of cancer, independently on an international level. Eventually,

he expanded its use to other illnesses. He was the director of the *Orgone Institute Research Laboratories* in Tel Aviv and the editor of the journal *Internationale Zeitschrift fur Orgonomie*, published during the early years of the 1950s.

6. Blasband, R.A.: *Orgonomic Functionalism in Problems of Atmospheric Circulation. Part Two - Drought*, Journal of Orgonomy, Vol. 4, No. 2, Orgonomic Publications Inc, New York, November 1970.

7. To evaluate the intensity of DOR, Baker created a *DOR-index* that basically makes simple numerical evaluation of several subjective atmospheric characteristics such as the presence of wind, humidity, the color of the sky, and the presence of clouds, as well as personal sensations of the climatic conditions. A sum-total of all the parameters is then added up. The bigger the value of the index, the greater the quantity of DOR present in the atmosphere. Additional information can be found in Rosenblum, F.C.: *The Electroscope III: Atmospheric Pulsation*, Journal of Orgonomy, Vol. 10, No. 1, Orgonomic Publications Inc, New York, May 1976.

8. *New York Times*, 28 August 1977, in Blasband, R.A.: *CORE Progress Report No. 8*, Journal of Orgonomy, Vol. 11, No. 2, Orgonomic Publications Inc, New York, November 1977.

9. 1 acre corresponds to about 0.4047 hectars.

10. DeMeo, J.: *Environmental Notes. Desertification*, Pulse of the Planet #2, Natural Energy Works, Ashland, Oregon (USA), Fall 1989.

11. Trenberth, K.: *Earth Dries Up As Temperatures Rise*, New Scientist, Vol. 185, No. 2483, 22 January 2005.

12. *New York Times*, 18 January 1975, in Blasband, R.A.: *Core Progress Report No. 5*, Journal of Orgonomy, Vol. 9, No. 1, Orgonomic Publications Inc, New York, May 1975.

13. Jenkins, G.: *The Scientific Basis for the Prediction of Impacts from Climate Change*, in Flannery, B.P., Kohlhase, K.R., LeVine, D.G., International Petroleum Industry Environmental Conservation Association Symposium on *Critical Issues in the Economics of Climate Change*, edited by Flannery, B.P. and Grezo, C.A., London, 1997.

14. *New York Times*, 21 August 1975, in Blasband, R.A.: *CORE Progress Report No. 7*, Journal of Orgonomy, Vol. 11, No. 1, Orgonomic Publications Inc, New York, May 1977.

15. Blasband, R.A., DeMeo, J., Morris, R.: *CORE Progress Report No. 15, Breaking the 1986 Drought in the Eastern US*, Journal of Orgonomy, Vol. 21, No. 1, Orgonomic Publications Inc, New York, May 1987.

16. Blasband, R.A.: *CORE Progress Report No. 4*, Journal of Orgonomy, Vol. 8, No. 1, Orgonomic Publications Inc, New York, May 1974.

17. Blasband, R.A.: *CORE Progress Report No. 2*, Journal of Orgonomy, Vol. 6, No. 1, Orgonomic Publications Inc, New York, May 1972.

18. Blasband, R.A.: *Orgonomic Functionalism in Problems of Atmospheric Circulation, Part Three. On Desert*, Journal of Orgonomy, Vol. 4, No. 2, Orgonomic Publications Inc, New York, November 1970.

19. Blasband, R.A.: *Orgonomic Functionalism in Problems of Atmospheric Circulation, Part One. The Normal Atmosphere*, Journal of Orgonomy, Vol. 3, No. 2, Orgonomic Publications Inc, New York, November 1969.

20. Blasband, R.A.: *CORE Progress Report No. 8*, Journal of Orgonomy, Vol. 11, No. 2, Orgonomic Publications Inc, New York, November 1977.

21. One degree Fahrenheit corresponds to 9/5 of a degree Celsius + 32. The variation of a degree Fahrenheit corresponds to 9/5 of the variation of a degree Celsius.

22. Blasband, R.A.: *CORE Progress Report No. 11, The Drought of 1980-1981*, Journal of Orgonomy, Vol. 15, No. 2, Orgonomic Publications Inc, New York, November 1981.

23. All data and maps were supplied by the *Weekly Weather and Crop Bulletin*.

24. Blasband, R.A.: *CORE Progress Report No. 13, Fighting the Extreme Drought of Spring 1985*, Journal of Orgonomy, Vol. 19, No. 2, Orgonomic Publications Inc, New York, November 1985.

25. *Weekly Weather and Crop Bulletin* (WWCB), USDA/USDOC/NOAA, Washington DC, 18 December 1984; 19 March 1985; 9 April 1985; 18 June 1985; 23 June 1985.

26. Schleining, J.: *CORE Progress Report No. 29, Summer Drought Relief in Southern Oregon. June 24 to July 9, 1992*, Journal of Orgonomy, Vol. 26, No. 2, Orgonomic Publications Inc, New York, autumn/winter 1992.

27. Blasband, R.A.: *OROP Hurricane Doria*, Journal of Orgonomy, Vol. 6, No. 1, Orgonomic Publications Inc, New York, May 1972.

28. Reich, W.: *OROP Hurricane Edna*, CORE, Vol. VII, No. 1-2, Orgone Institute Press, Rangeley, Maine (USA), March 1955.

29. One mph corresponds to 1.609 km/h.

30. Eden, J.: *Operation Firebreak*, Journal of Orgonomy, Vol. 13, No. 2, Orgonomic Publications Inc, New York, November 1979. Also see Eden, J.: *CORE Manual, Cosmic Orgone Engineering*, Careywood, Idaho (USA), 1986.

31. Schleining, J.: *CORE Progress Report No. 17, Fighting Forest Fires and Breaking the 1987 Drought in the Northwest US*, Journal of Orgonomy, Vol. 22, No. 1, Orgonomic Publications Inc, New York, May 1988.

32. Eden, J.: *OROP Schweitzer Basin*, Journal of Orgonomy, Vol. 13, No. 1, Orgonomic Publications Inc, New York, May 1979.

Healing of Atmospheres

Cloudbusting in Recent Times
The Experiments of James DeMeo

"... South Sudan Rains Turn Famine Zone Green. The rains have finally arrived in Bahr el Ghazal, the area of South Sudan worst affected by famine, transforming the region from an unforgiving dustbowl into what looks like an oasis ...

... Some flooding also reported along the Nile river in Sudan in the Khartoum lowlands. Lake Nasser, behind the Aswan Dam in Egypt, is also at very high levels due to the good rains, now flowing into Nile River tributaries."

9 September 1998, Reuters CNN Internet

"News Report of an excellent grain harvest in Eritrea ... the best crop in 40 years ...."

2 November 1998, "Eritrea Has Unusual Farm Aid", Associated Press, AOL Internet

One of the biggest proponents of orgonomy and in particular of *cloudbusting* is currently James DeMeo, Director of the *Orgone Biophysical Research Laboratory*[1]. Since the mid-1970s, DeMeo undertook a series of experiments, both locally and extended over vast areas, to verify the effectiveness of the technology on both atmospheric processes and climate. His results have been resounding. Not only has *cloudbusting* been demonstrated to be valid in producing rain and restoring, at least *temporarily*, the natural atmospheric behavior, but DeMeo has demonstrated that it even works in extreme climatic conditions and environments, typical of arid and desert areas, such as exist in the Southwestern arid-zone of the United States, and in sub-Saharan and southern Africa.

DeMeo's work is without doubt monumental and of enormous importance, because he was the first university scientist to experimentally and scientifically verify, through extensive field tests and a systematic evaluation of the results, the original ideas of Reich.

In addition, the meteorological parameters related to the operations have always been supplied by official sources or agencies (such as the National Weather Service – NWS – for the experiments carried out in the United States) and taken from weather stations close to the intervention areas. The results obtained have always been compared with the previous conditions of the area, when these were available. In the final analysis, the development of the atmosphere and its perturbations have also always been monitored, both before and during DeMeo's operations, through satellite images.

The evolution of his research and experimental work has brought him to develop a more dynamic and interactive *cloudbusting* than that of Reich, who acted individually and in an essentially local way. DeMeo confirmed what Reich noted firstly, that very often operations carried out from a single position do not have good results, since the natural atmospheric flow can be blocked by a barrier of stagnant air situated in an area that can be a long distance from the intervention area. Following Reich's example, he found that it was necessary in the majority of cases to intervene from several positions, and occasionally to use a larger number of *cloudbusters*. In most cases, he used just one *cloudbuster*, assembled on a trailer, which was continually moved from one position to another, but eventually he used up to five different cloudbusters, both large and small, usually separated by many miles distance and operated in a coordinated manner.

The first systematic studies conducted by DeMeo were undertaken towards the end of the 1970s through field tests, carried out in collaboration with the Geography-Meteorology Department at the University of Kansas, where he worked as an instructor and eventually earned a doctorate degree. His experimentation included 12 tests, both to increase the dimension of the clouds and to produce rain. These were carried out under normal climatic conditions.

Subsequently, he undertook field experiments that were progressively more extensive and under more harsh climatic conditions, in areas characterized by medium to high levels of drought, aridity or desert conditions. DeMeo carried out numerous operations in the United States[2], in Kansas and later on when employed as professor of geography at Illinois State University and the University of Miami, and finally through his private institute[1]. In the 1990s, he directed experiments on semiarid and arid lands in southern Europe[3] and in the arid and desert areas of Africa[4]. The results were positive in all these experiments, and in nearly all cases they were better than expected. His field work eliminated the droughts and brought significant rains, even to nearby desert areas that were not worked with the *cloudbuster*. His later experiments were undertaken only during extreme droughts, or in desert regions, with no forecast of rains. He nevertheless obtained a high percentage for the stimulation of rain and the recovery of the natural atmospheric functions.

Even during an operation in Israel during extreme drought, the rains which followed were so abundant it was necessary to carry out an intervention to reduce the rains, which were threatening to create damages to parts of the region.

Mostly, DeMeo performed *cloudbusting* operations aimed at reducing drought or the aridity of a determined area. In one case, reported in the literature[5], with the help of a collaborator, Robert Morris, he undertook a successful operation to slow down and halt the southerly movement of a polar cold air mass which was moving into Florida from the north, threatening to freeze the citrus crops in the area.

Some years later, he directed a *cloudbuster* experiment in northern Germany, towards reduction of air pollution and restoring health to dying forests[6].

During all his activities, over thirty years worth, DeMeo has always worked with a close and trusted group of serious and professional collaborators. Among these were the Americans Richard Blasband, Robert Morris, Theirrie Cook, Matthew Ryan and Donald Bill; the Germans Bernd Senf and Stefan Müschenich; the Italians Carlo Albini and Aurelio Albini; and in Greece, Georgos Argyreas; also there were many other temporary helpers and volunteers. DeMeo has also carried out a series of interventions in the USA, starting from 1990, with help from the *USA CORE Network*. Within the *CORE Network*, information on the weather is exchanged, the climatic conditions are discussed, possible interventions in certain areas are decided, and results critically reviewed with safety oversight.

One of the firm points in DeMeo's research activities is the safety during the use of the apparatus. He continually repeats the warnings on the dangers both to the operators and to the surrounding environment, should the equipment be used in an inaccurate or irresponsible way. DeMeo, like Reich and other researchers, tries to make the operators aware and responsible for their operations in order to avoid any damage to their own organisms or the environment. In a *booklet*[7], he wrote of numerous people who, applying *cloudbusting* in an "activist" and incompetent way, only for curiosity, sense of potency or profit, or with deeply mystical motivations, had created or intensified hard droughts or forest-fires, or created violent cloudbursts, causing great environmental and property-damage in the areas where they were operating, sometimes with avoidable deaths of people.

Another area in which DeMeo has always placed much importance is that of understanding the dynamics of deserts and droughts around the planet. Based on his own observations and field experience, he has elaborated a very interesting theory on the development of drought and desertification on a global level[8]. He has identified an area of the planet as a unique and interconnected desert and semi-arid region, which he calls *Saharasia*. It is a wide band, situated in the Northern Hemisphere, which covers the driest and harshest land and marine areas, mainly across North Africa, through the Middle East, and into Central Asia: *Sahar-asia*. This region plays a fundamental role in the atmospheric processes and in the formation of the climate even in areas situated far away from it. According to DeMeo, Saharasia has firstly been the cradle where the armored and desert character structure of human beings originated. It is secondly the largest and most extremely dry zone on the planet, and out of which very hazy-dusty dry air and DOR spreads out to infect other world regions, to form what he calls *secondary deserts*, which are found in the most diverse regions of the planet. In Figure 17, one can see the extension of this band of territory across extreme-arid Saharasia, with long extensions out towards other semi-arid regions of the planet. The desert areas are shown in black while the areas that have dry seasons at one or another time of year are shown in grey. DeMeo observed that droughts always result when the overheated and dry DOR air covering Saharasia pushes outwards into bordering wet or merely semi-arid regions, drying them out unexpectedly for long periods.

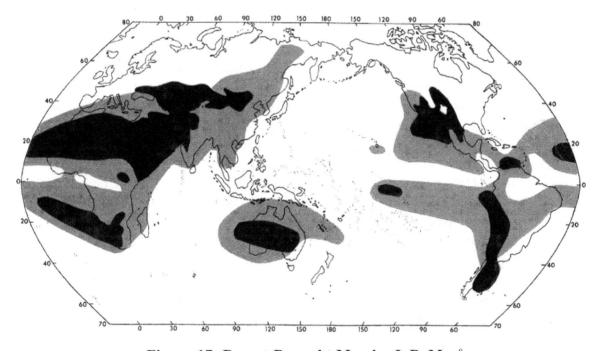

Figure 17: Desert-Drought Map by J. DeMeo[8]

DeMeo observed that this wide region seems to be at the center of the global process of desertification. It contains a great mass of stagnant atmosphere (DOR) which can extent outwards to affect other parts of the planet. It seems that this stagnant atmosphere is spread all over the globe by the planetary winds, creating secondary-deserts and droughts. Some of these are for example the *Santa Ana* and the *Chinook* in the United States, which are typically very hot and dry. These blow at very high velocities and cause irritation and illness as well as drought, heat waves and fires in the areas that they cross. In other areas of the world, these winds have different names like *Foehn*, *Sirocco*, *Hamsin*, *Sukhoevi*, and *Wind of 120 Days*, but they essentially have the same effects on the environment and the people. They are also called *bad winds* and can blow from the desert towards areas situated even at long distances away. They can retain their dry characteristics unaltered, even after having travelled for thousands of kilometers over the open oceans or through temperate and humid areas. In this way, dry winds are characterized by stagnant DOR air which comes from the Sahara and can cause drought in distant areas such as Florida and South America. If the stagnant air lingers over a certain area, it can start fires with the loss of wood and forest lands. It is this first step that can bring about the formation of arid and desert conditions in those areas.

By DeMeo's Saharasian theory, deserts are seen as a general source of arid and dry climates even in regions that are far away and are normally humid and temperate.

Using Reich's bioenergetic concepts of water-hungry stagnant orgone (DOR), DeMeo compared the deserts to large thirsty ameba[9] which send out pseudopods of stagnant air or DOR towards more humid territories, both near and far.

It is therefore understandable that the deserts of drylands are the origin of the oceanic deserts. In turn, these can feed stagnant atmospheric conditions over neighboring lands, creating heat waves and forest-fires there as well, and thereby adding to the overall warming of the planet.

Often atmospheric *dry fog*, which originates from stagnant DOR-containing air blowing out of the deserts over the open ocean, has been observed over the open ocean and along coastlines of the North Atlantic, Indian Ocean, Persian Gulf, West-Coast USA, and elsewhere.

As DeMeo noted, it is as if Earth suffers from a great injury or tumor, which continues to develop and grow. This tumor manifests itself in hyperarid and desert territories of the Saharasia belt, but its expansive action gives rise to new secondary-desert areas, in the same way a tumor can metastasize.

The dramatic conclusion of this global process is the progressive advancement of the desert that is currently hitting our planet, expanding at a rate of 60,000 square kilometers each year. According to DeMeo, this process has been going on since 4000 B.C., when the desert band first started to form. All the documents analyzed have the same conclusion, that before 4000 B.C. the climatic conditions of this wide area, today covered by hyper-arid desert conditions, was then completely different and character-ized by a humid climate with trees, grasslands, all kinds of animals and large populations[10]. The reasons for the formation of this giant desert area are still unknown.

On a practical level, DeMeo's first experiences with this phenomenon was the presence of a barrier of stagnant air (DOR) during one of the operations carried out in Kansas in the period of 1979-80[11]. Despite repeated interventions carried out in Kansas, stagnant DOR-infested dry air always moved back into the region, coming from the southwestern deserts of the USA. He was only temporarily able to stimulate the atmosphere and produce rains in this situation.

Everytime the atmosphere seemed to give rise to a cloud system, it quickly disappeared, frustrating the efforts. After weeks of work and experiments, he had the impression that the obstacle was not to be found in Kansas, but in the dry atmosphere of the desert region off to the southwest. Considering the observations and studies made by Blasband some years earlier, he recognized that the arid area of the Central US was in some way connected with distant areas. These areas were situated both to the west and southwest of North America, where the orgone energy streams and major Westerly winds normally enter the continent.

DeMeo noted that the flow of one of the main energy currents, the Galactic Orgone Stream, suddenly stopped and disappeared from the satellite images. Its course was then diverted towards the north, along the Rocky Mountains. The deviation seemed to be driven by an invisible barrier that the Galactic Stream was hitting. In this manner, it inhibited the current's flow over the central part of the USA, favoring the spreading of a stagnant atmosphere into this area, which came directly from the southwestern deserts. The initial effect of the stagnant air was that of immobilizing the atmosphere, giving rise to drought that could persist for weeks or even months. In addition, it impeded the flow of new energy streams within

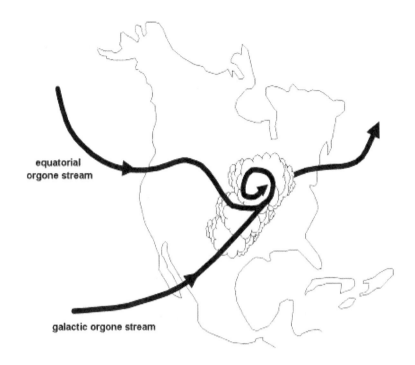

Figure 18: Superimposition of Energy Streams Under Normal, Healthy Atmospheric Conditions. (from DeMeo[11, 12])

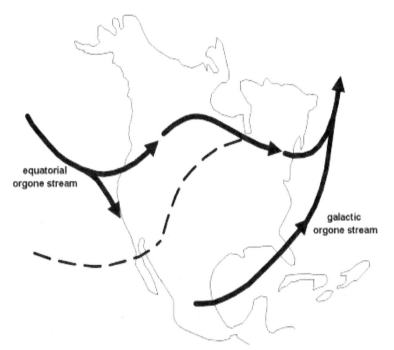

Figure 19: Split Jet Stream and Block Under Abnormal, Dor-Drought Atmospheric Conditions. The Diagram Below Gives a Schematic Representation of the Two Conditions. (from DeMeo[11, 12])

Outbreak of DOR desert air into a moist border region

Orgone prevails
DOR is sequestered in the moister region as storms develop, sometimes with severe activities

DOR prevails
and is not sequestered; DOR spreads, anchors itself to the landscape, produces drought, diverts all streams of energy and moisture

the area.

Under normal conditions, the energy currents responsible for the usual atmospheric conditions of the United States are the Equatorial Stream and the Galactic Stream. Usually the Galactic Stream enters North America via western Mexico with a flow from the southwest towards the northeast. Meanwhile the Equatorial Stream flows from west to east. From the superimposition of these two energy streams and moisture-carrying wind belts, the summer rains and the winter cyclones develop over the central agricultural regions of the United States known as the *Great Plains* and the *Midwest*.

Figure 18 schematically shows the direction of the flow of the two energy streams and the central area of the United States in which they meet[11, 12].

During the period of drought, neither of these two streams penetrate the central part of the USA. The Equatorial Stream is diverted towards the north, above the Canadian border, while the Galactic Stream is blocked and deflected from its normal course towards the areas of Arizona and New Mexico.

At the same time occurs the formation of atmospheric processes in the central part of the United States which are characterized by stagnant air and elevated pressures. This situation was responsible for the drought that hit the area at the end of 1980. Figure 19 schematically illustrates the situation. The broken line in the Figure indicates the flow of the Equatorial Stream still in a normal direction, but very weakened.

More generally, it can be said that the start of the stagnant desert air in a determined area, where temperate and humid climatic conditions are present, can be functionally represented by the diagram in Figure 19[13].

As one can see, in addition to the formation of stagnant air in humid areas, it can also happen when the area is sufficiently charged with dynamic orgone energy, that the DOR is not able to prevail and is therefore isolated. In this case, the area reacts with the formation of strong storms that develop along the borderline between the humid area and the stagnant air that comes from the desert.

Cloudbusting in the USA

The first experiments carried out by DeMeo were in the second half of the 1970s, during his period of research at the Department of Geography-Meteorology, University of Kansas[14,15]. The experiments were planned to test the effects of the *cloudbuster* on individual clouds, under steady weather conditions. Some of the most important parameters monitored during these experiments were the size and the dynamics of the clouds. The *cloudbuster* was pointed towards a cloud for a short period. The evolution of the cloud was then followed and monitored over time. The cloud was photographed at regular time intervals, both under normal conditions before the experiment, then during and after the influence of the *cloudbuster*. The analysis of the photographs revealed a general reduction in the dimensions and the speed of growth of the clouds following the pointing of the tubes.

Figure 20 shows one such example, of the trend in the area of a cloud both before and during the action of the *cloudbuster*. It can be very clearly seen that the development of the clouds was inhibited by the *cloudbuster's* action.

In the same period, DeMeo conducted 12 experiments with the *cloudbuster* in Kansas to produce or increase the quantity of rain under normal climatic conditions[16]. In these experiments, the *cloudbuster* was put into action for a period of only a few hours. Figure 21 summarizes the results of the data analysis which covers a period of 7 days. The zero identifies the day when the operations were carried out. The three days on the left of the graph indicate the total rainfall in Kansas prior to the intervention, and those on the right the total rainfall after the intervention. All the data was supplied by the *National Weather Service* (NWS) on a sample of 278 weather stations spread across the region. The marked increase in

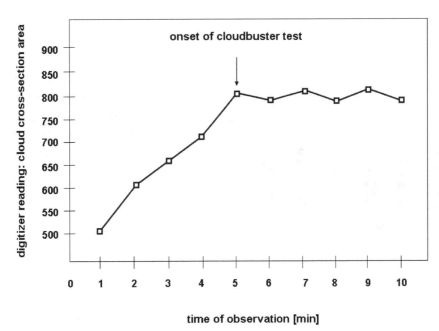

Figure 20: Inhibition of Cloud Growth via Cloudbusting (from DeMeo[14, 15])

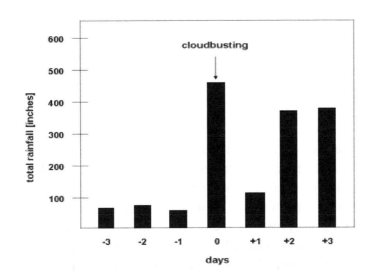

Figure 21: Daily Precipitation Data for 12 Cloudbusting Operations in Kansas, Encompassing Three Days Before Operations, the Day of Operations (at Zero), and Three Days After Operations. Data from 278 weather stations were used for this graphic. (from DeMeo[16])

rainfall following the intervention is obvious from the graph.

Seventy-five percent of the operations were also characterized by a reduction in the atmospheric pressure, which indicated an increased probability of rain. In fact, in nine cases (75%) out of the 12 considered, rainfall occurred within 48 hours.

Later, between 1979 and 1980, another 12 operations to produce rain were undertaken in the northeast of Kansas[17]. These were aimed at bringing rain to areas with a medium-high drought level,

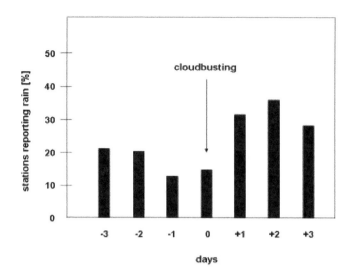

Figure 22: Change in Percent of Weather Stations in Kansas, Nebraska, Iowa and Missouri Reporting Rain, for the Period of 12 Cloudbusting Operations in Kansas. (from DeMeo[17])

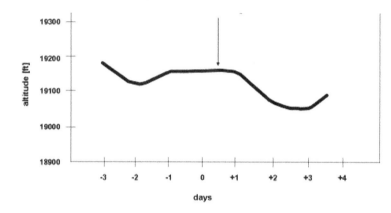

Figure 23: Change in Altitude of the 500 mb Pressure Surface, Before, During and After Cloudbusting Operations in Kansas. (from DeMeo[17])

characterized by a marked absence of rain in the area over weeks or months. In addition, in each case the weather forecast gave the probability of rain as zero or almost zero on the days of and immediately following the interventions. In this group of experiments, 10 of the 12 operations produced rain in the region surrounding the *cloudbuster* within 6 to 42 hours from the start of the intervention.

DeMeo saw that the operations performed in areas subjected to drought could bring rain within 48 hours, while those carried out in areas with normal climatic conditions (like those of Kansas during the 1977-78 period) could produce rain within 24 hours. This is because very often *cloudbusting*, when undertaken in a very dry area, requires preliminary maneuvers to mobilize the atmosphere which could last for some additional hours.

Figure 22 shows the trend of the weather stations percentage that recorded daily rainfall in Kansas, Nebraska, Iowa and Missouri during the 12 experiments carried out in northeastern Kansas in 1979-1980, for a period of 7 days. An increase in the percentage of the weather stations that registered rainfall after the interventions is evident.

Figure 23 shows the trend of the average altitude of the 500 mbar pressure on northeastern Kansas, in the period both before and after the interventions. It can be noted that the altitude of the 500 mbar pressure decreasing at about 100 feet[18] coincided with the first day after the interventions, with a maximum depression occurring within 36 to 48 hours thereafter.

In the period between 1981 and 1982, DeMeo carried out two operations in Florida, planned for the 30th and 31st of December 1981, and the 18th and 19th of March 1982[19]. The aim was to combat the heavy drought that had affected Florida and all of the southeast United States since 1980. All the operations

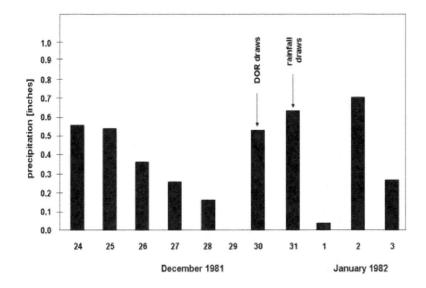

Figure 24: Precipitation Data, Florida Operations, Dec. 1981 to Jan. 1982. (from DeMeo[19])

Figure 25: Precipitation Data, Florida Operations, March 1982. (from DeMeo[19])

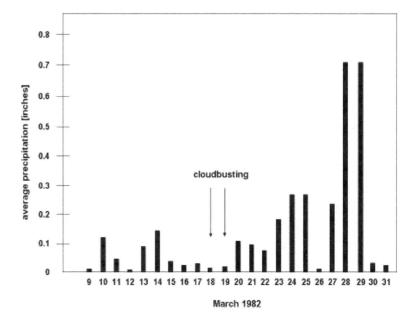

were carried out with just one mobile *cloudbuster* which was moved from one location to another within the drought zone, so as to influence three or four different locations in one day. The operations from each single position generally did not last for more than one hour.

The operations of the first phase were characterized both by the absorption to eliminate DOR and mobilize the atmosphere, and by absorptions aimed at contracting the atmosphere to produce rain.

Immediately after the start of the operation, a low-pressure system developed over the Gulf of Mexico, connected to a warm front extended towards the east to where the *cloudbuster* had been operating. No fronts were present on the satellite maps taken before the intervention, and the NWS weather forecast had not predicted rain in those days. The perturbation continued to gradually develop as the operations proceeded, until a great storm system finally moved into the operational area. Figure 24 shows the rainfall data for this first phase.

The persistent drought in Florida made a second phase of the operation necessary, which had already been planned for the 18th and 19th of March 1982. According to the weather forecast, was going to be no rain in that time period. Contrary to this, rainfall increased gradually after the interventions, both in Florida and the entire southeastern United States. During the following two weeks, abundant rain fell

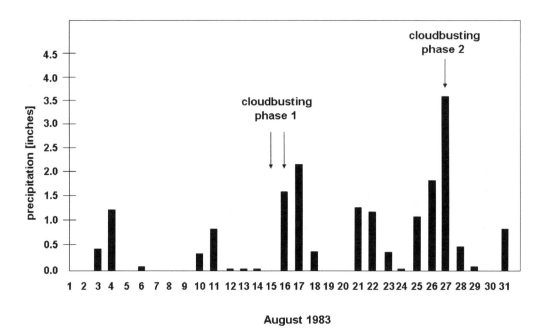

**Figure 26: Precipitation Data, Illinois Operations, August 1983
(from DeMeo[20])**

on the whole area. Figure 25 shows the average rainfall trend both before and after the second phase of the operations (March 1982).

The following year, DeMeo undertook two experiments in Illinois, aimed at ending a strong drought[20]. This operation was subdivided into two phases; the first was carried out on the 15th and 16th of August, and the second on August 27th. Both the phases were accompanied by storms across the whole area.

Figure 26 shows the daily rainfall trend for Illinois (from 12 weather stations) for the whole month of August. The large peaks in the rain that was recorded immediately after the intervention are notable.

DeMeo found that around 79% of the interventions that he had performed in the period from 1977 to 1983 in Kansas, Florida and Illinois, had produced rain within 48 hours after the start of the operation. This agreed with what Blasband[21] had established based on his observations. In fact, in order for an intervention to be considered effective, rain must be produced within 8 to 48 hours from its start, on a day when the probability of rain was less than 10%.

In 1988, facing one of the worst drought that had hit the central and northwestern parts of the United States, also characterized by extremely high temperatures, DeMeo organized two series of long-range interventions which affected vast area of the USA, from the Gulf of Mexico to the Canadian border[22]. The first operation had its center in Arizona, while the center of the second one was along a line between Seattle and Spokane, in the State of Washington.

The first phase was set up in such a way as to bring rain in the desert of the Southwest, which in turn would stimulate storm systems moving northeast towards the central parts of the US.

DeMeo thought that this could be accomplished by restoring the normal flow of the Galactic Stream over the central parts of the USA, which was found to be off course during that period. This movement would then attract a southward motion of the Equatorial Stream that was situated over Canada, therefore bringing rain to the central parts of USA as it would happen under normal conditions, thanks to the superimposition of the two main streams (see Figures 18 and 19).

The operations of the first phase started on August 12th and ended on August 14th. The operative plan

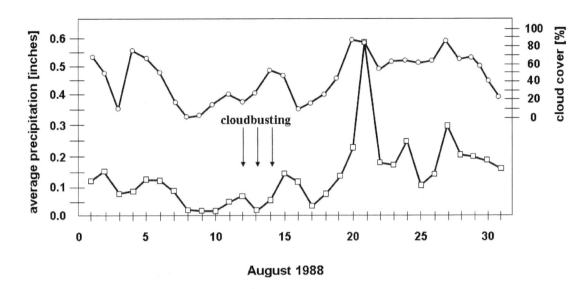

**Figure 27: Precipitation and Percent Cloud Cover Data,
Arizona Operations, August 1988 (from DeMeo[22])**

included an extensive absorption aimed at cleaning the atmosphere of DOR and reducing the stagnated atmosphere. Successive absorptions were aimed at stimulating the flow of the Galactic Stream from southwest towards the northeast and attracting the Equatorial Stream southward. While the NWS weather forecast for the 14th of August did not predict any rain in the area of the *cloudbuster's* position, satellite images showed big clouds and storms throughout the whole area. The following morning, a great system of humid air, coming from the southwest, moved into the region, bringing rain to the whole area and into the arid central USA. Figure 27 shows the average rainfall trend (from 177 weather stations) and the percentage of cloud cover during the month of August 1988, for the state of Arizona.

Figure 27 shows peaks both in the cloud cover and in the rainfall after August 13th and 14th, days that were characterized by long absorption operations. An increase in the cloud cover and in the rainfall occurred even in the following week, with a strong increase starting on August 17th. In addition, the city of Yuma, situated a short way to the south of the *cloudbuster's* position, recorded more than 200% of the normal rainfall for the month of August, despite the weather forecast not predicting rain for that period. There was an increase in the rainfall and a decrease in the temperatures even in the central parts of the United States. This fact substantially confirmed DeMeo's hypothesis, and Reich's overall theory, on the necessity of superimposition between the Galactic and Equatorial Orgone Streams for the production of abundant rain in the central parts of the US.

The second phase of the operation concerned the Pacific Northwest of the United States consisting of Northern California, Oregon, Washington, Idaho, Montana, and Wyoming, which did not benefit from the rainfall that was generated in the previous phase. The temperatures continued to rise, and forest fires devastated the area. In July of 1988, fires destroyed 1.5 million acres, while around 3 million acres had been burned since the beginning of that year. By the end of August, there were about 66,000 individual fires burning in the region.

The operative program drafted by DeMeo and his group included absorptions to eliminate the DOR. These were mostly carried out from Seattle and eastward into Spokane (Washington), using a mobile *cloudbuster* unit. He planned long absorptions to be carried out, aimed at breaking the barrier of stagnant air in the desert, and bringing humidity from the coast to inland regions, and thereby producing rain across the Pacific Northwest.

All the minimal details were planned out, and when a last check was performed on the equipment during the transportation of the *cloudbuster* from the base in San Francisco (site of the first phase of the operation) towards Seattle (Washington), a strange phenomenon was noticed. Intense hail and strong gusts of wind occurred in the area where the *cloudbuster* was parked. These phenomena were surprising, because the equipment had neither been grounded in water nor the tubes extended.

Operations were established firstly in Seattle, and continued towards Spokane, starting on the morning of September 13th and ending on the morning of September 16th.

The operations produced intense cloud cover and persistent rain, lightning and occasional hail, the first change in the weather for months in that area. In addition, the fires in the Washington area were being reduced and extinguished, and even snow had fallen in small quantities. The rain continued to fall in the following days both on the coast and inland. *Cloudbusting* had effectively eliminated the barrier of stagnant air that had opposed the flow of humid air from the Pacific inland and had prevented the formation of storm systems. Additionally, a mass of clouds and strong gusts of wind developed directly along the direction the *cloudbuster* was being towed to the east by the truck. The next evening in Spokane, intensive clouds formed surrounding the parking area and it rained for the first time in months that night, an event which was not forecasted by the NWS. The following day, with further work, the rains spread east from the Pacific Ocean across Washington and Idaho, towards Montana and Wyoming, where a strong snowstorm occurred over Yellowstone Park. A dramatic forest fire of historical proportions in Yellowstone Park was unexpectedly ended with the unforecast snowfall and rain, which at the very least had been intensified by the *cloudbusting* operations.

During these operations, the operators were subjected to an energy overcharge due to the enormous quantity of *Oranur* produced. The excess was reduced and kept under control through known remedies, such as prolonged water bathing.

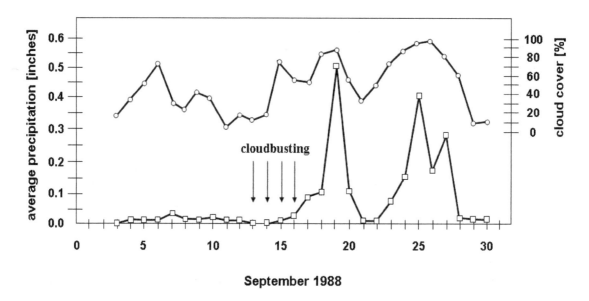

**Figure 28: Precipitation and Percent Cloud Cover Data for
Washington State, Pacific Northwest Operations,
September 1988 (from DeMeo[22])**

Figure 28 shows the percentage of cloud cover and the average rainfall (from 150 weather stations) in the state of Washington for all of September 1988.

As can be seen from the graph, the quantity of rain that fell before the start of the operations was practically negligible when compared to the amount coming down after the interventions.

Aside from the very successful nature of this operation, DeMeo thought that the strange local weather reactions, when the *cloudbuster* was not even set up and working, were caused by the physical movement of the *cloudbuster* from one area to another[23], which excited the atmosphere.

These phenomena had already been observed by McCullough, Reich's assistant, in October 1954, during the moving of a large truck-mounted *cloudbuster*, from Maine towards Arizona[24]. He noted a strong excitement in the atmosphere during the transport, even though its tubes were pointed low and the end of the tubes was capped. He saw the development of clouds to the west and southwest of the direction the truck was being moved.

DeMeo had also observed similar phenomena sometimes previously during the moving of the *cloudbuster* from one location to another. In January 1988, during the move from Florida to Iowa, rain and strong gusts of wind accompanied the movement of the truck. He discovered that with the slowing or stopping of the vehicle, the intensity of the storm diminished. After having slowed the vehicle to 25mph, the wind calmed down and the storm diminished in its intensity.

Again, on the 27th of May 1988, while DeMeo was moving his *cloudbuster* from Iowa to Arizona, there was an unforeseen storm on the arid area of Nebraska, where he stopped for the night. This storm brought 3 or 4 inches of rain and hail stones of half an inch in diameter, gusts of wind and many broken branches. The strange thing was that the heavy rain, the hail and the wind were restricted to an area of half a mile around the place where the equipment was parked.

DeMeo suggests that these phenomena must be taken under serious consideration. When the possibility of creating strong storms during the transportation of a *cloudbuster* can occur, it is necessary to slow down or stop it completely.

In 1989, DeMeo undertook an ambitious research project with the goal of studying the long-term effects of *cloudbusting* on the climate and vegetation of the deserts of the southwestern United States[25,26,27,28]. At the same time, the project would have also cast light on the geographic expansion of these effects. The operation consisted of five short interventions on pre-announced dates, each separated from the other by a period of three and a half weeks during which all operative activities were suspended.

After a preliminary study on the atmospheric behavior of the area, the project was started in the spring of 1989. A mobile *cloudbuster* was transported to and set up in the most arid region of the southwestern desert, along the Colorado River, around 100 km north from Mexico on the border between California and Arizona. Figure 30 highlights the position of the *cloudbuster*. This position allowed easy access to and guaranteed abundant water for the operations. The equipment was left in position for 3 to 6 days only. A second *cloudbuster* was placed in the area close to San Francisco Bay in California, operated on occasion by Blasband. This was done on two different occasions to support the principal operations in the desert. At the end of each phase, the group left the area, taking with them the mobile *cloudbuster*. This allowed the climatic conditions to recover and stabilize. In addition, it allowed the group to carry out analysis of the data and, based on the results, to plan the next operative phases.

The first phase of the operation started during the second week of May 1989 and lasted three days, from the 8th to the 10th of May. The temperature in the area surrounding the principal position during the first day was 100-105°F. The working hypothesis basically was to induce the convergence of the Galactic and Equatorial Orgone Streams into the area of the *cloudbuster*.

When the two energy streams moved inland under the influence of the instrument, the first from northwest and the second from the south, they gradually gave rise to abundant rains and strong gusts of wind which spread towards the north into Montana. The first phase ended on the evening of May 10th. That day the temperature of the area had reached a maximum of only around 70°F.

Figure 29 shows the average rainfall related to the first phase of the intervention. The data supplied

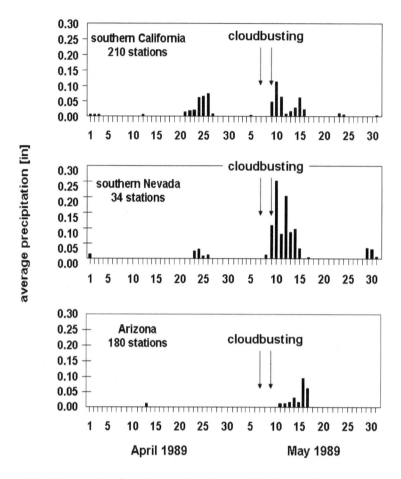

Figure 29: Precipitation Data for First OROP Arizona Experiment, May 1989 (from DeMeo[25, 26, 27, 28])

Figure 30: Study Region for OROP Arizona Experiments, 1989 (from DeMeo[25, 26, 27, 28])

by NWS for the months of April and May 1989 came from 424 weather stations.

Figure 30 shows the geographical distribution of the 424 weather stations used to collect data. Of these, 210 stations were located in southern California, 34 in southern Nevada, and 180 within the state of Arizona. From Figure 29, the increase in the rainfall following the intervention is evident. A great intensity of rain can also be clearly noted for Southern California and Nevada. The time between the start of the operations and the first rainfall was shorter (about one day) for California and a little longer (around 4 days) in the arid and desert regions of Arizona.

The second phase of the operations took place from June 5th to 8th, 1989. This phase was characterized by high temperatures and an immobile stagnant atmosphere. After two days, no great changes had been observed, but there had been a slight reduction in the air temperature. It was decided to carry out an absorption with the second *cloudbuster* positioned in the area of San Francisco, with the aim of mobilizing the atmosphere. Shortly after the intervention, a low-pressure storm system moved from the southeast of San Francisco Bay, where the second *cloudbuster* was, towards the position of the main *cloudbuster* in Arizona. This storm system caused abundant rainfall on the southern mountains of the Sierra Nevada, an improbable event in the area at that time of the year. The operation was ended immediately after this rain.

The third phase took place from July 3rd to 5th, 1989. The temperature was 110°F with relative humidity under 15%. After one day of operations, some large storms developed to the south of the principal position in northwestern Mexico, but no storms penetrated to the Arizona area. Also on the second day of the operation, temperatures touched 123°F. This operation was terminated on the third day with high temperature and pressure. A final absorption was attempted, combining the *cloudbusters* in Arizona and San Francisco. There were no short-term results, but the following week it rained on the entire area.

The fourth phase was carried out between 31 July and 3 August 1989. The operation started with a situation that was already quite good. Almost daily, storms and rain were recorded for the area. The objective this time was to increase the rain beyond the average normal quantity.

After the first day of operations, a wide system of storms developed to the east of the main position. At the end of the second day, the relative humidity had risen to 70%. This operation ended without any support from the other *cloudbuster*. Yuma, Arizona, subsequently had a rainfall equal to two years of normal conditions, while other regions of southern California recorded abundant rains. Some areas of the Great Basin deserts had high quantities of rain. By the end of August, the area close to the main *cloudbuster* already had rain equal to 500% of their monthly normal value. The last phase of the operation took place between the 25th of August and the 1st of September 1989. After one day of operations, partially cloudy conditions developed over the banks of the Colorado River. Sparse rain was recorded across Arizona, particularly in the mountains. A wide storm system developed in Mexico, to the south of Yuma, and in Arizona. The main *cloudbuster* was then moved from its usual position westwards to the coast of California, where the tubes were grounded in the waters of the Pacific Ocean. The operations were concluded after half a day of absorptions without appreciable results. Nevertheless, the tropical depression that was found in the South Pacific, many miles away, moved towards the coast of Mexico. Between the 15th and the 17th of September, an unusual and abundant rain was brought to the area of southern California, the first significant rainfall in two years for many locations.

Figure 31 shows the data of the average daily rainfall for the 424 weather stations, related to the whole period of the operation: from April to October 1989. The arrows on the graph correspond to the intervention phases.

The results of this systematically-undertaken experimental cycle, with dates planned and announced to the NWS in advance, were without doubt positive. The collected data indicated that four of the five operations were followed by significant rain, not attributable to chance or pre-existing atmospheric

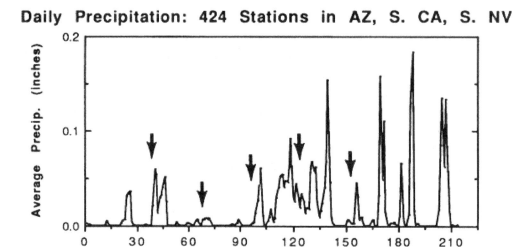

Figure 31: Precipitation Data for Five OROP Arizona Experiments, 1989
(from DeMeo[25, 26, 27, 28])

conditions. It had a success rate of 80%, very close to that already observed in previous studies.

The data for the average daily rainfall, related to the 424 weather stations, was then transformed into a percentage for the 30 days, calculated with respect to the maximum rainfall value. This calculation was carried out for each of the five phases. The data was then superimposed and averaged. Figure 32 shows the trend of this quantitative analysis. The zero of the graph corresponds to the first day of the intervention, while the 15 days to the left correspond to the rainfall values of the two weeks prior to the interventions. The 15 days after the interventions are reported to the right of the zero. This approach shows in a more highlighted way the close relationship between the interventions and the rainfall in the considered area.

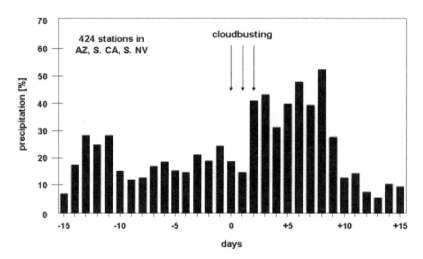

Figure 32: Average Percent Precipitation Data for 15-Day Intervals Before
and After Operations, for Five OROP Arizona Experiments, 1989
(from DeMeo[25, 26, 27, 28])

From Figure 32 it can be seen that the quantity of rain that fell after the operations is about double the amount which had fallen prior to the interventions. As can be noted, the increase in the rainfall developed around 48 hours after the start of the operation and lasted for about a week.

DeMeo continued with his work, and two interventions were carried out between the end of March and the start of April 1990, in central California and the Sierra Mountains, at the invitation of local farmers[29]. The rainfall of the last months had been very low and drought was threatening the local orchard crops. The rainfall following the operation was more than 400% of the normal quantity for the area to the east of the position of the *cloudbuster*, while in the region to the south values around 150% were recorded[30]. In addition, in the month of May abundant rain fell on a large area surrounding the operative area. The rainfall varied from 200% to 400% of the normal value.

Towards the end of May 1990, an expedition to the far northeastern part of Montana was undertaken at the request of another group of local farmers[31]. The intention was to bring a flow of humidity to the area which had been subjected to a long and heavy drought. The operation started around the end of May and lasted for a week. The absorptions were carried out from different positions across the states of Washington, Idaho, Montana, and North Dakota.

48 hours after the start of the operations, a large storm system formed in the northeast of Montana, which brought sparse rain into the whole region. In some areas, it was the first rain in two years. The following week brought the second and third largest rainfalls out of that entire summer's record. Nevertheless, as had already occurred many times before, from mid-July on the rain in the northeast of Montana had almost ceased and the drought intensified once again. A second expedition into that region was not organized due to a lack of funds, and because one of the farmers thought he could easily "do it himself" – he was not successful.

After the experience in Montana, DeMeo carried out other interventions in California[32], from October 1990 to April 1991, working closely with Richard Blasband. California was then in a continuous fight against drought and desertification, the origin of which, according to DeMeo, are to be identified in a large region of desert-like conditions in the nearby Pacific Ocean which dries out the coast of southern California and northwestern Mexico. By DeMeo's analysis, this desert ocean condition extends all the way from the Saharasian deserts, into the Indian Ocean, and from there into the Pacific Ocean and finally spreading into the Southwestern USA. Ending severe droughts in California therefore requires to address this phenomenon.

The operations performed during that period were able to bring rain to the whole area, particularly to the northern parts of California. Figure 33 shows the values of the daily rainfall[33] for the whole period of the operation, from October 1990 until April 1991. The horizontal rectangles show the days on which *cloudbusting* was carried out from the principal position situated to the north of San Francisco. The operations were characterized by a combination of short absorptions aimed at eliminating the DOR and long absorptions intended to bring rain. The vertical rectangles show the days on which the main *cloudbuster* was used at the same time as others localized in different positions, both in California and Oregon.

The direct correspondence between the operations and the daily rainfall is evident from the graph, especially when the interventions were carried out at more than one position. One fact of particular interest is the rainfall recorded for February and March of 1991, using more than one *cloudbuster*. This phase was called *Miracle March* by the weather forecasters and newspapers, due to the enormous quantity of water that fell following the interventions. Other than the principal *cloudbuster,* which was situated to the north San Francisco and operated by Blasband, another four *cloudbusters* were used. One of these was a large mobile *cloudbuster*, which was operated by DeMeo from multiple locations across southern to northern California. The other cloudbusters were small devices worked by collaborators, from locations in southern Oregon, and in the *Central Valley* and *Sierras* of California. The operations were coordinated by telephone, and planned based on satellite images. The team was composed of nearly a dozen operators from the *USA Core Network*, distributed in five different locations. The first

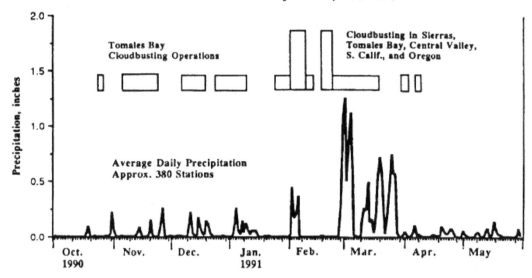

Figure 33: Precipitation Data, California Experiments, 1990-1991 (from DeMeo[32])

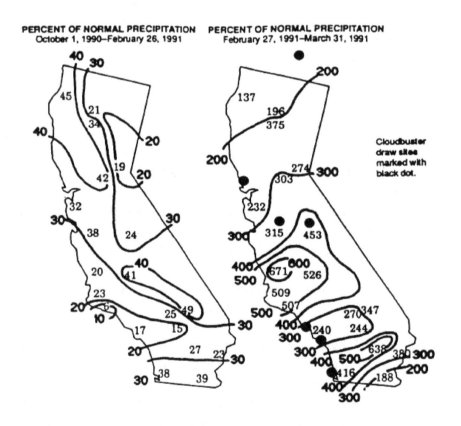

Figure 34: Percent of Normal Precipitation Maps, Before and After California "Miracle March" Experiments, 1990-1991 (from DeMeo[32])

phase using multiple cloudbusters was started on February 1st and ended on February 8th. During these first days, 0.75 to 1.5 inches of rain fell on *Central Valley* in California. In addition, 10 to 18 inches of snow fell in areas near the *cloudbuster* sites, both in Oregon and in parts of central California. An operator of the mobile *cloudbuster* became ill despite the precautions undertaken, because of the high concentrations of toxic energy created around the position. He was treated with methods known to be of benefit.

After this, a second expedition took place starting on February 18th, once again using all the *cloudbusters*. It was terminated on February 21st, leaving just one *cloudbuster* working occasionally. During those days, an enormous storm front approaching the Californian coast was created, and this finally removed the barriers which had created the drought. It started to rain on the 27th of February and didn't stop for 4 days, covering the whole state, and this was followed by a series of additional storms. *Cloudbusting* operations were continued from the one California location until the first week of April with the goal of keeping the weather moving into the area. Figure 34 shows the percentage values with respect to the normal values for the periods prior to (from October 1990 to February 1991) and after the combined interventions in the state of California (March 1991). The black circles indicate the positions of the *cloudbusters*. California newspapers called these the "Miracle March" rains, which saved the state from a severe water crisis. But all efforts to gain the attention or help from state authorities, both before and after the operations, received no notice or reaction whatsoever. This magnificent result was greeted with total silence, except among the friends and workers of the *Core Network*.

Cloudbusting in Europe and Israel

At the start of the 1990s, DeMeo was invited by various working groups in Europe, to apply the *cloudbusting* methods to critically-severe atmospheric conditions, particularly in Greece and Israel.

In 1990, on the invitation of the *Hellenic Orgonomic Association*, he carried out several interventions in Greece. These were aimed at fighting a severe and persistent drought that was occurring in both Greece and a large part of Italy, Turkey, and the Balkans[34]. DeMeo again theorized that the dry conditions were the consequence of an abnormal influence of the Sahara Desert atmosphere. One operation was carried out at the end of May. At that time, the water reserves of Athens were only going last for 40 more days. The interventions directed by DeMeo went on for two days, using a large hydraulically controlled *cloudbuster* constructed by Georgos Argyreas, with help from the Greek organization. A week of rain followed those operations which covered all of Greece, as well as parts of Italy and Turkey. A successive operation was performed in summer, at the start of the dry season, but it did not produce significant results, except a reduction in the temperature and the DOR levels.

Finally, a third operation was set up in November of the same year. This final intervention used several *cloudbusters*, positioned in various areas of Greece, Crete and Cyprus. The use of all the instruments was coordinated under DeMeo's instructions by telephone from the Thessaloniki station, the site of the main *cloudbuster*. After three days of operations, a large storm system developed to the west of Gibraltar that started to move eastwards, bringing heavy rains to a large part of the Northern Mediterranean and abundant snow both to the Alps and the Balkans, ending the drought that had gripped the Mediterranean.

In 1991 DeMeo was invited to carry out interventions in Israel, with permissions and logistical support from the Israeli government, to work against the remaining parts of the drought that had hit the whole of the Mediterranean[35], but which had not benefited from the prior year's work in Greece.

Starting from 1988, in a large part of the Mediterranean, the rainy season had been below average. Few cyclonic systems had entered the area in those years and all the nations in that part were suffering from a lack of water. In Israel this brought about a progressive reduction in the water reserves and a lowering of the level of the most important lakes, which were the main water resources of the whole country. In January 1991, the level of Lake Kinneret (Sea of Galilee) was at its historical lowest for 60 years. The Lake supplies about a third of the water needed for agricultural and domestic use in Israel, while the remaining part comes from water wells that access the groundwater tables.

A feasibility study undertaken by DeMeo showed that the possibility of restoring natural atmospheric pulsation and rains over the area was quite high, especially following the earlier success in Greece. Experiments had also already been carried out in Israel during the 1960s and 1970s by Hoppe and his associates[36, 37, 38] with very positive results, and later by Blasband[39] in 1983. Also significant were the 1963/64 tests carried out by the Israeli meteorologist Gad Assaf, who observed significant decreases in the 500 mbar pressure surface over the region after *cloudbusting*.

With help from Theirrie Cook, DeMeo organized an intervention to restore rains to Israel, divided in two phases: the first in November 1991, and the second in February 1992. The second phase was planned to prolong any possible benefits from the first phase. In addition, if the November operations were a failure, the second phase would be a second chance to stimulate rain.

A small but powerful *cloudbuster* was built which could be maneuvered manually and easily loaded onto a truck for transportation (Photo 6, top). In the days prior to the first phase, Israeli television announced that Lake Kinneret was at its historical lowest level, being only two millimeters above the emergency "red line" due to drought and heavy water demands. The operations were started on November 15th and concluded on November 24th. Additional smaller supporting *cloudbusters* were also occasionally operated during this same period at DeMeo's instructions in Greece and Cyprus. Everything was coordinated by telephone. It started to rain significantly in Greece and southern Italy on the 17th of November, with a slow progressive eastward movement of the storm fronts. In Israel, the rain started to fall on the 27th of November and continued for all the following week. Persistent rain was also registered in Lebanon, Cyprus and Turkey. The rainfall in that period, from November 27th to

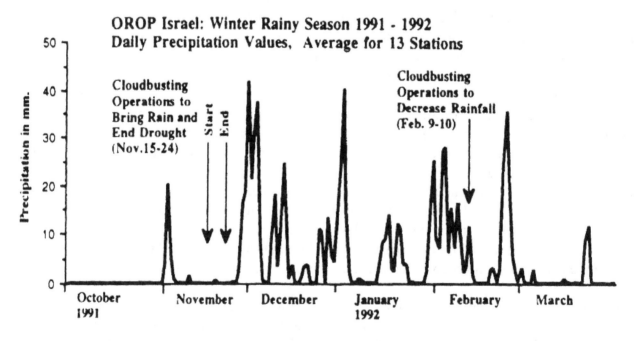

Figure 35: Precipitation Data, Israel Experiments, 1991-1992
(from DeMeo[35])

Photo 6 – top: Cloudbuster *Sabra* at Work on the Shore of Lake Kinneret (Sea of Galilee), Tiberius, 17 November 1991. Bottom: Lake Kinneret in Early February 1992, After Several Months of Good Rains. (Courtesy of James DeMeo)

December 7th, brought to Israel and its neighboring territories a rain equal to 200-400% of the normal amounts for that period. A 50-year rainfall record was broken in Israel. The mountains of Turkey, Cyprus, Syria, and Lebanon registered unexpected and abundant snow. Lake Kinneret rose to 60 cm above the emergency level by December 15th. Jerusalem registered a record of 16 inches of snow while in Amman, Jordan, 24 inches fell. In the following months of January and February, the abundant rains over the entire eastern Mediterranean made the second phase of work unnecessary. DeMeo described the situation as follows[40]:

> *"Indeed, in my 15+ years of working with the cloudbuster in drought and desert regions on three continents, I had never before witnessed such a powerful response to cloudbusting – the impression gained was of the sudden release of an incredible, accumulated atmospheric tension, with an associated shift of climate back towards what probably existed prior to the original desertification of the Saharasian region several thousand years ago."*[40]

Unfortunately, during the entire rainy period, the Israeli government also continued with a program of non-stop *cloudseeding*, which DeMeo felt was unnecessary[41] and warned them against. As a consequence, too much rain fell in some regions and caused some damage, mostly because the storm-drains and diversion canals had not been maintained properly. However, traffic accidents actually declined, possibly due to people slowing down because of the poor visibility and the flows of water that ran down some of the roads. The strong winds and the abundant snow left a few areas without power for some days. Snow fell even on some parts of the Negev desert, which had never before happened in living memory. The enormous quantity of water falling on the countryside created numerous small lakes which also had never been seen previously. Lake Kinneret continued to accumulate water and actually threatened to flood the coastal cities situated around the Lake, particularly the town of Tiberius. To prevent this, the authorities decided to open the Deganya Dam to allow excess water to flow from the Lake into the Jordan River, down into the Dead Sea, something which also had never happened before. Water was also pumped back into groundwater reservoirs. By the end of March, no flooding had occurred, the Lake was completely full, and groundwater reservoirs were significantly recharged.

Photo 6 (bottom) shows Lake Kinneret in early February 1992, after several months of good rain. Notice the increased height of the water level as compared to the condition in late November 1991 (top).

The agriculture benefited from this abundant rain and a higher production was recorded. The abundance of fruit and vegetables was so high that within a few months, the prices on the Israeli market had dropped by 14.3%, which caused a record drop of 0.4% in the cost of living, after 23 years[42].

The *Jerusalem Post* from 15 February 1992 commented the situation as follows:

> *"It's official. The current and far-from-over winter is the wettest in a century in Jerusalem and Tel Aviv, where there hasn't been so much rain since records were first kept in 1904. 75 cm of snow on Mt. Hermon, the heaviest snowfall in 140 years. In the Golan Heights, snow is considered a blessing... Mt. Hermon is composed of eroded limestone, which absorbs water into its many crevices. The melting snow trickles down into multi-year reservoirs, which feed the headwaters of the Jordan river, providing the only source of water to the Kinneret during the summer."*

It should also be mentioned, an operation was undertaken in February of that year, to slow down and divert the *Jet Stream* flow, which was continually bringing rains to the eastern Mediterranean. This intervention was carried out at DeMeo's request from a position in Greece, and appeared successful in halting the excess rains created due to the non-stop *cloudseeding* activity which DeMeo had warned against. Nevertheless, overall the damages caused by excess rains were very small in comparison to what had been created previously by the multi-year hard drought conditions. The exceptional rains and snow produced a great bounty and blessing for the region.

Figure 35 shows the daily average rainfall for 13 weather stations in Israel, for the period from 1 October 1991 to 31 March 1992.[43] The graph displays both the periods of the intervention for the production of rain and the second short intervention carried out to reduce the rainfall in February.

Cloudbusting in Africa: Namibia and Eritrea

The most amazing and interesting experiments conducted by DeMeo over his 30 years of work, were the expeditions into the arid and inaccessible deserts of Africa, even if they sometimes were the most dangerous. In the last ten years of the 20th century, he led multiple expeditions into two African countries, into the Namib and Kalahari deserts of Namibia, and secondly into the desert areas of Eritrea, at the southeastern edge of the Sahara Desert.

Between 1992 and 1993, with collaborators from the USA (Theirrie Cook, Donald Bill), Germany (Manfred Metz, Bernd Senf), and Namibia (Fritz Jeske), DeMeo led two separate expeditions into Namibia.[44,45] At that time, Namibia, South Africa, Botswana, and Zimbabwe were all overcome by drought and facing food shortages affecting millions of people.

After various months of study and planning, it was decided to carry out two expeditions. The first, in

Photo 7 – OROP Namibia, February 1993. James DeMeo at Hardap Dam.
(Courtesy of Bernd Senf)

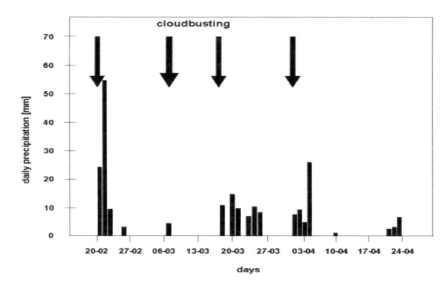

**Figure 36: Precipitation Data, Omaruru Namibia,
February-March-April 1993 (from DeMeo[44])**

November 1992, was substantially dedicated to the building and preliminary testing of the apparatus that would be used in the second expedition. The operations were carried out on the second expedition, in February and March 1993 (Photo 7), to coincide with the onset of the normal rainy season, although natural rains had been greatly diminished over many years with a 12-year drought, being most acute in the last 3 years. By contrast, the 1993 *cloudbusting* operations were followed by a significant amount of precipitation. In addition, over the following months of March and April, there was a noticeable amount of increased rain on all of southern Africa.

While a full analysis of precipitation data for the Namibia experiments was never undertaken (as DeMeo reported to the author, he was blocked by jealous and upset Namibian government meteorologists from obtaining a full set of weather data), some information was gathered from local newspapers. Figure 36 shows the trend of daily rainfall for a weather station situated in the vicinity of the *cloudbuster*, 60 km northeast of Omaruru. From the graph one can see that there is a close correlation between the dates of operations and episodes of rainfall.

A 5-Year Desert-Greening Project in Eritrea

The *Green Sea Eritrea* project without doubt confirmed more than any other the effectiveness of *cloudbusting* when applied to produce rain and end prolonged drought in arid and desert areas of the Earth. DeMeo led a team which included collaborators from Europe (Carlo Albini, Aurelio Albini, Bernd Senf), the United States (Theirrie Cook), and Eritrea, who carried out interventions across Eritrea, at the southeastern border of the Sahara Desert.[46] Two separate expeditions were carried out each summer over five years, starting in summer 1994, in some cases using up to three large *cloudbusters* at the same time. These were situated at different points of the country. The project was organized with the full support of the Eritrean government and the financial support of their Agricultural Ministry.

Eritrea had suffered through a period of almost thirty years of drought which had nearly reduced the entire nation into an arid and desert land. In this situation, the Eritrean government had to spend around $100 million or more each year to import food. The *cloudbusting* method offered a hope to return the natural rains, increase the agriculture and local production, and reduce the imports. The total cost of the operation for the whole 5 years was approximately $65,000, as paid to DeMeo's non-profit institute. While the initial operations were financed by donors from the USA and Europe, subsequent work was funded by the Eritrean government after officials saw the positive results of the first year. This was the first time in the history of orgonomy that any government had financed a *cloudbusting* project.

The project planning began in 1992 with an in-depth study of the region's climate conditions, of logistics and landscape, and a search for financial sponsors for the first year of work. In the higher plains of Eritrea, the rainy season goes from June to September and is characterized by a northward movement of the Inter-Tropical Convergence Zone (ITCZ), a belt of rainy weather and Tropical Easterly winds which covers most of the Equatorial regions. In the months between October and May, there is the dry season with nearly no rainfall, as the Sahara Desert atmosphere then covers the entire region.

In the last 30 years, the ITCZ and Tropical Easterly winds, which are responsible for the rains on the higher plains, rarely moved northwards, leaving the entire region under the influence of the Saharan atmosphere. This happened because of the presence of thick levels of DOR that formed in those years over the whole region, reaching high up into the atmosphere, upwards to 6,000 meters above the ground level. Under these conditions, the ITCZ and Tropical Easterlies remained far to the south, except on sporadic occasions. To bring rain and humidity to the higher plains, and generally to the whole of Eritrea, it was necessary to remove the stagnant atmospheric block, to bring the ITCZ and Easterlies back northward, and also to stimulate a related flow of humid air which naturally comes towards Ethiopia and Eritrea from the southwest, from the Gulf of Guinea in West Africa.

The operative program included the positioning of the main *cloudbuster* in strategic points of the region, so as to effectively absorb and remove the DOR. Successive absorptions were aimed at bringing humidity to the region from the Indian Ocean, the Red Sea and the Gulf of Guinea.

The *cloudbuster* was placed each time close to artificial lakes, created from dams made during the period of the Italian colonization. The operative group was made up of 3 to 6 people. The government of Eritrea supplied logistic and operational support, a translator with truck and driver, as well as the materials and the manpower needed for the construction of the necessary devices.

In addition, the *Research Division* of the Ministry of Agriculture and the *Meteorological Analysis Branch* of the Civil Aviation Department, lent their support for the collection of meteorological data and satellite images, and for access to various points of the territory.

Once built, the *cloudbuster* (shown in Photos 8 and 9) was transported and set up along the shores of the lakes chosen for the operation, at a certain distance from the base camp, and made to work for short periods. Absorptions from just one site for many days were avoided, preferring to carry out shorter interventions from different locations. The most applied technique was the DOR-*busting* in those areas that seemed most charged with DOR, accompanied by absorptions aimed at contracting the orgone energy envelope, to produce rain. The meteorological data was collected from the weather station at the Asmara airport. However, more frequently weather maps and other information was sent via fax from the USA, and supplemented by information gathered from short-wave radio.

The work was carried out in two separate phases. The start of the first phase coincided with the start of the rainy season (so as to ensure a good quantity of natural rainfall), while the second phase started halfway through the rainy season (to ensure the continuation of the rain). It was planned on the following days:

• phase I: 20 June - 3 July 1994
• phase II: 18 August - 27 August 1994

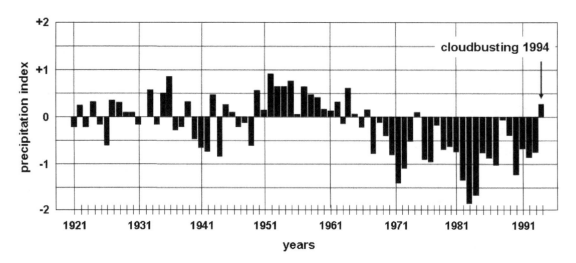

Figure 37: Precipitation Index for the African Sahel, 1920-1994
(from DeMeo[46])

The operations of that year culminated with abundant rainfall on all of Eritrea and the area of the Red Sea, which filled numerous dry lakes. In addition, the rivers were also filled, benefiting nearby countries such as the Sudan and Ethiopia. The rain was so abundant that in one situation, the position of the *cloudbuster* was almost put in danger. It was parked on the top of a small hill, in the middle of a dry lake. The rain fell so quickly that the lake filled within a very short time and threatened to transform the hill into a small island in the middle of the lake. Only a fast rescue operation under pouring rain, in the middle of the night, prevented this from happening.

Figure 37 shows the Precipitation Index for the area of the Sahel over the many years before the *cloudbusting* operations.[47] As can be seen, the annual index of the rainfall reveals the 30 years of drought in the Sahel, but shows an inversion in 1994, the first year of the *cloudbusting* operations.

In the summer of 1995, two expeditions were performed which again gave excellent results. The *cloudbusting* was carried out from the same positions as the year before, and took place in the following periods:

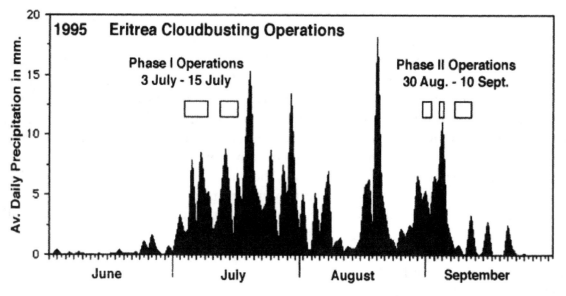

Figure 38: Precipitation Data, Eritrea Experiment 1995
(from DeMeo[46])

• phase I: 3 July - 15 July 1995
• phase II: 30 August - 10 September 1995

As in the previous year, the operations were followed by long and abundant rains that covered most of Eritrea, with benefits to nearby Ethiopia and Sudan. Dry lakes across the region were often filled beyond capacity. In particular, by the end of July, the major reservoir Lake Mainifhi had risen by 3 to 5 meters from its original level.

Figure 38 shows the average daily rainfall data for Eritrea for the months of June through September of 1995, from a network of 39 Eritrean government weather stations[48]. The rectangles indicate the periods in which the *cloudbuster* was active.

From the Figure, a close correlation between the large quantity of rain that fell and the start of the operation can be observed. It must be noted, however, that after the start of September there were only isolated storms. This was due to the Saharan atmosphere that once again started to push south, limiting the effectiveness of the operations.

Despite the positive results obtained in the period of 1994/95, the operation for 1996 was cancelled when the Eritrea government and the Ministry of Agriculture withdrew support from the project, after several critics had argued that the project lacked scientific validity and was a waste of public money.

Consequently, after two good years of results, there was no *cloudbusting* undertaken in 1996, and rain was scarce that year. Figure 39 shows the trend of the average daily rainfall on Eritrea in 1996 from the same 39 weather stations[49]. The complete absence of peaks in the rainfall that characterized the good rain of 1995 can be noted.

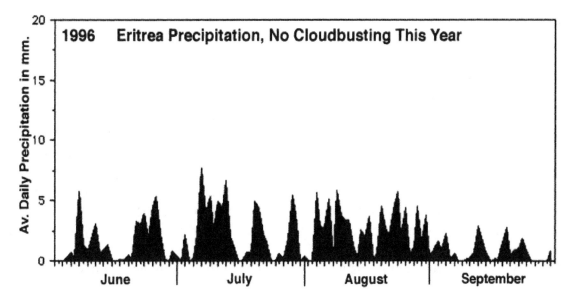

Figure 39: Precipitation Data, Eritrea 1996. No Cloudbusting.
(from DeMeo[46])

As a consequence, the Eritrean government was faced with an increase in the costs of food in that year, which had been drastically reduced following the abundant harvest of the period 1994/95, resulting from abundant rainfall following the *cloudbusting* operations. The import of food during 1994/95 when *cloudbusting* was performed was about 30 to 50 million dollars per year, while in 1996 the cost of food imports rose to approximately 150 million dollars.

Because of this fact, the Eritrean government contacted DeMeo again in the spring of 1997 and

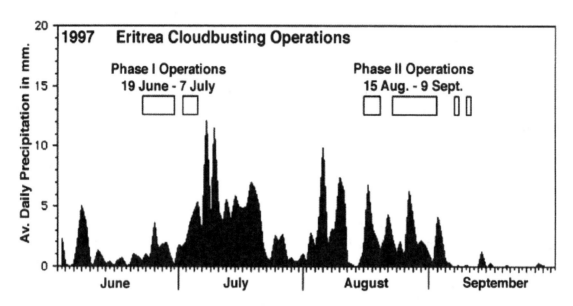

Figure 40: Precipitation Data, Eritrea Experiment 1997
(from DeMeo[46])

requested that he and the team return once again. With some guarantees that the original 5-year project would be completed, operations resumed in Summer 1997.

The 1997 operations took place in two distinctive phases, following the same criteria as before:

- phase I: 19 June - 7 July 1997
- phase II: 15 August - 9 September 1997

The same techniques were applied as in 1994 and 1995. Figure 40 shows the trend of the average daily rainfall on Eritrea from the 39 weather stations[50].

The high peaks of rain can be noted on the Figure, especially in the first phase of the operation. The *cloudbusting* in the second phase seemed to have a lesser effect, probably due to the effect of the *El Niño*[51]. DeMeo considered that the use of just one *cloudbuster* could be insufficient for operations under particular atmospheric conditions, like in the presence of the *El Niño* which also correlates with dry conditions over the Indian Ocean and East Africa. Under those especially difficult conditions, one instrument appeared insufficient to keep the atmosphere alive and moving. He decided for the operation of the following year, to work with more than one *cloudbuster*, as had been done many times previously during other operations.

In the fourth year, 1998, the operations were carried out in the following two distinctive phases:

- phase I: 15 July - 7 August 1998
- phase II: 12 August - 31 August 1998

This operation was conducted under very difficult climatic and social conditions, due to a restart of the conflict between Ethiopia and Eritrea in May of that year. The operations proceeded nevertheless, this time using three *cloudbusters* which were placed in climatically strategic locations and coordinated via short-wave radio. The new *cloudbusters*, completely built in Eritrea, were similar in energetic drawing power to the main one, but much more primitive in construction. Use of the three *cloudbusters* in tandem produced a powerful result, even though this was restricted to no more than a few hours of work per day, and not more than a few days of work from any one location.

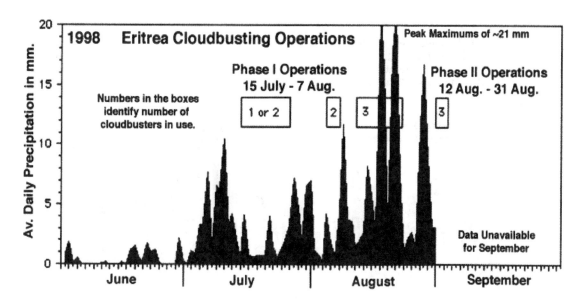

Figure 41: Precipitation Data, Eritrea Experiment 1998
(from DeMeo[46])

Figure 41 shows the trend of the average daily rainfall on Eritrea in 1998 from the 39 weather stations[52]. Once again, the presence of wide peaks of rainfall can be observed corresponding to the operations. The rectangles indicate the days of work and the number of *cloudbusters* used.

As can be seen from Figure 41, the first phase of the operation did not produce much rain, but this was when only one or two of the *cloudbusters* were being used, and the system for coordination of operations was being tested. DeMeo found it difficult to synchronize the operations from many sites, due to a lack of trained and experienced people, but this problem was eventually overcome. In the second phase he decided to use a maximum of two *cloudbusters* at the same time and for not more than two days from any one location, and then only at night. The results, as can be seen from the Figure, were spectacular. Three wide peaks of rain occurred during this second phase. The rain falling between the 15th and the 23rd of August registered as the highest average daily rainfall that had been achieved during the entire period of the operations. The fields of the farmers were saturated, the harvest increased in an exceptional way, and the lakes were completely full. In addition, abundant rain was also recorded in the neighboring states of Ethiopia, Sudan, Yemen and Saudi Arabia. The rains of 1998 appeared to be the most plentiful so far.

Notwithstanding these positive results, the subsequent operations of 1999 were reduced, with the presence of just one operator in Eritrea, and with loss of financial support and with limited logistical support from the Ministry of Agriculture. All this was due to the return of the war conditions between Ethiopia and Eritrea. Asmara airport was continually attacked by Ethiopian jets and repeatedly bombed. The military conflict took place mostly along the southwest border between the two countries, with sporadic incursions inland. Only the first phase of operations was planned and carried out, given the potential dangers. In the subsequent months, the conflict intensified until it reached the most inner regions of Eritrea. Hence the project was suspended and then definitively cancelled. While Eritrean officials and news reports indicated 1999 was another very good year of rains, DeMeo was not able to obtain rainfall data for that year.

Figure 42 shows the results of the analysis carried out on the data of 1995, 1997 and 1998. The diagram shows the fraction of the average daily rainfall, as percentage referred to the maximum daily rainfall, for the fifteen days before and after the start of the operations.

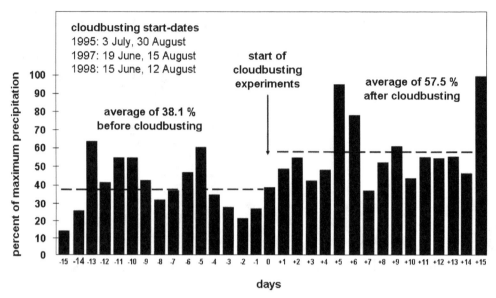

Figure 42: Average Percent Precipitation Data, Eritrea Experiments, 1995, 1997 & 1998 Combined, for Days Before and After Operations (from DeMeo[46])

The marked increase in the rainfall in the fifteen days after the operations is evident from the graph. It passes from an average value of 38.1% before operations, to 57.5% after the operations. The overall rainfall, in the period after the *cloudbusting*, represents over 50% more than the overall rainfall relative to the previous period. A t-test of probability of the rainfall and the distribution shows a value of p<0.0042, which means that these could have happened by chance only four times out of every thousand.

As a testimony to the success of the 5-years of work in Eritrea, some newspapers[53] commented in December 2000 on the unprecedented formation of immense new lakes in the middle of the Sahara Desert, some hundreds of kilometers north of Eritrea, but downstream within the Nile River Valley. These events followed the exceptional rainfall which occurred during the years of the Eritrean *cloudbusting* operations, which always benefited the surrounding areas (see Figure 43). These were the consequences of the enormous quantity of water flowing into the River Nile which slowly filled Lake Nasser for the first time since it had been created, following the construction of the Aswan High Dam in 1968. The excess water which flowed into the Nile River was carried into Lake Nasser, which filled to capacity. The overflow was then directed from the Lake out into the open Sahara Desert, where it created giant new lakes, a phenomenon that has never previously been observed.

Besides the media, also NASA was interested in the event, documenting it very clearly with satellite pictures[54] and commenting on the formation of new lakes in the middle of the Sahara as a quite uncommon occurrence:

"Four lakes formed recently in southern Egypt in an area that was previously desert. Fed by unusually high levels of rainfall and water overflowing from the Aswan High Dam on the Nile River, the first lake appeared in 1998. The Aswan's overflowing waters are channeled through the arroyo into a reservoir, as expected, but as the high rains have continued, so has the overflow. Consequently, the reservoir has grown in size and three more lakes have formed.

.... Scientists don't know whether or not the lakes will remain, or will dry up within a few years."

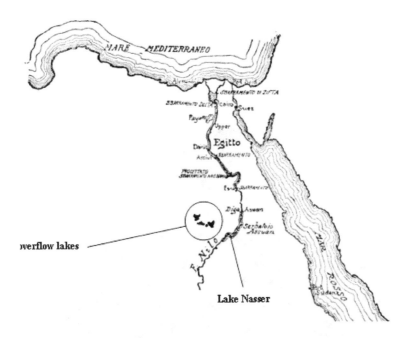

Figure 43: Overflow Lakes Near Aswan Dam & Lake Nasser, Egypt, Downstream of Eritrea Operations Which Affected the Upper Nile River Basin[46]

Photo 8 – James DeMeo With Cloudbuster *Kiremti* After Several Days of Work in the Bush, August 1998. Other Major Participants in this Successful 5-Year Project Included: Theirrie Cook, Aurelio Albini, Carlo Albini, and Bernd Senf. (*Green Sea Eritrea* Operation, Courtesy of James DeMeo)

Photo 9 – Cloudbuster *Kiremti* Standing on a Small Hill Next to the Mainifhi Reservoir, Near Asmara, Just After a Heavy Rain, August 1994. Reproduced in Color on Rear Cover. (*Green Sea Eritrea* Operations, Courtesy of James DeMeo)

Chapter 6 Citations

1. The *Orgone Biophysical Research Laboratory*, founded in 1978, has its office in Ashland, Southern Oregon. It has the goal of promoting and spreading scientific research in the field of sex-economy and orgone biophysics. Research activities and experiments are particularly concentrated on cross-cultural sex-economic investigations, laboratory tests aimed at the study of the applications of orgone energy, and the use of the orgone accumulator and the *cloudbuster*. It publishes the journal *Pulse of the Planet* and books on the research activities of the Institute. http://www.orgonelab.org

2. The interventions in the United States were carried out in Florida (with 2 interventions), Illinois (2), Georgia and South Carolina (3), Arizona (4), the State of Washington (4), California (5), and the arid desert area of the Southwest (6).

3. The interventions were carried out in Greece (7), Cyprus (7), and Israel (8).

4. The interventions were carried out in Namibia (9) and Eritrea (10).

5. DeMeo, J., Morris, R.: *CORE Progress Report No. 14, Possible Slowing and Warming of an Arctic Air Mass Through Cloudbusting*, Journal of Orgonomy, Vol. 20, No. 1, Orgonomic Publications Inc, New York, May 1986.

6. DeMeo, J.: *Orgonomic Project Waldheilung 1989-1993*, Pulse of the Planet #4, Ashland, Oregon (USA), 1993. Also see DeMeo J.: *Addendum to CORE Progress Report No. 15, Reduction of Rainwater Acidity Following the End of the 1986 Drought: An Effect of Cloudbusting?* Journal of Orgonomy, Vol. 21, No. 1, Orgonomic Publications Inc, New York, November 1987.

7. DeMeo, J.: *So You Want to Build a Cloudbuster? On the Problem of Growing Interest in Cloudbusting. A Personal View.* Orgone Biophysical Research Laboratory, Greensprings Center, Ashland, Oregon (USA), 1989. Revised Edition 2002, http://www.orgonelab.org/sobuildaclb.htm. Also see: http://www.orgonelab.org/chemtrails.htm

8. DeMeo, J.: *Desert Expansion and Drought: Environmental Crisis, Part I*, Journal of Orgonomy, Vol. 23, No. 1, Orgonomic Publications Inc, New York, May 1989. Also see DeMeo, J.: *"Desert-Drought Map"*, Pulse of the Planet #2, p.82, 1989. Also the chapter on "The Saharasian Desert Belt" in DeMeo, J.: *Saharasia: The 4000 BCE Origins of Child-Abuse, Sex-Repression, Warfare and Social Violence, in the Desert of the Old World*, Natural Energy Works, Ashland, Oregon (USA), 1998.

9. DeMeo, J.: *Desert Expansion and Drought: Environmental Crisis, Part I*, Journal of Orgonomy, Vol. 23, No. 1, Orgonomic Publications Inc, New York, May 1989. Also see DeMeo, J.: *"Desert-Drought Map"*, Pulse of the Planet #2, p.82, 1989.

10. DeMeo, J.: personal communications, December 2002.

11. DeMeo, J.: *Core Progress Report No. 20, Breaking the Drought Barriers in the Southwest and Northwest U.S.,* Journal of Orgonomy, Vol. 23, No. 1, Orgonomic Publications Inc, New York, May 1989.

12. DeMeo, J.: *Desert Expansion and Drought: Environmental Crisis, Part I*, Journal of Orgonomy, Vol. 23, No. 1, Orgonomic Publications Inc, New York, May 1989. Also see DeMeo, J.: *"Desert-Drought Map"*, Pulse of the Planet #2, p.82, 1989.

13. DeMeo, J.: *Desert Expansion and Drought: Environmental Crisis, Part I*, Journal of Orgonomy, Vol. 23, No. 1, Orgonomic Publications Inc, New York, May 1989. Also see DeMeo, J.: *"Desert-Drought Map"*, Pulse of the Planet #2, p.82, 1989.

14. DeMeo, J.: *Preliminary Analysis of Changes in Kansas Weather Coincidental to Experimental Operations with a Reich Cloudbuster*, M.A. Thesis, University of Kansas, Geography-Meteorology Dept., Lawrence, 1979.

15. DeMeo, J.: *Field Experiments with the Reich Cloudbuster: 1977-1983*, Journal of Orgonomy, Vol. 19, No. 1, Orgonomic Publications Inc, New York, May 1985.

16. DeMeo, J., 1985, *ibid.*

17. DeMeo, J., 1985, *ibid.*

18. One foot corresponds to 0.3048 meters.

19. DeMeo, J.: *Field Experiments with the Reich Cloudbuster: 1977-1983*, Journal of Orgonomy, Vol .19, No. 1, Orgonomic Publications Inc, New York, May 1985.

20. DeMeo, J., 1985, *ibid.*

21. Blasband, R.A.: *Orgonomic Functionalism in Problems of Atmospheric Circulation. Part Two - Drought*, Journal of Orgonomy, Vol. 4, No. 2, Orgonomic Publications Inc, New York, November 1970.

22. DeMeo, J.: *Core Progress Report No. 20, Breaking the Drought Barriers in the Southwest and Northwest U.S.*, Journal of Orgonomy, Vol. 23, No. 1, Orgonomic Publications Inc, New York, May 1989. Also see DeMeo, J.: *Cosmic Orgone Engineering Report*, Pulse of the Planet #2, Natural Energy Works, Ashland, Oregon (USA), Fall 1989.

23. DeMeo, J.: *Core Progress Report No. 20, Breaking the Drought Barriers in the Southwest and Northwest U.S.*, Journal of Orgonomy, Vol. 23, No. 1, Orgonomic Publications Inc, New York, May 1989.

24. Reich, W.: *Contact with Space*, Core Pilot Press, New York, 1957.

25. Blasband, R., DeMeo, J.: *CORE Progress Report No.21*, Journal of Orgonomy, Vol. 23, No. 2, Orgonomic Publications Inc, New York, November 1989.

26. DeMeo, J.: *CORE Progress Report No. 22, OROP Arizona 1989: Part II*, Journal of Orgonomy, Vol. 24, No. 1, Orgonomic Publications Inc, New York, May 1990.

27. DeMeo, J.: *CORE Progress Report No. 23, OROP Arizona 1989: Part III*, Journal of Orgonomy, Vol. 24, No. 2, Orgonomic Publications Inc, New York, November 1990.

28. DeMeo, J.: *OROP Arizona 1989: A Cloudbusting Experiment to Bring Rains in the Desert Southwest*, Pulse of the Planet #3, Natural Energy Works, Ashland, Oregon (USA), 1991.

29. DeMeo, J.: *CORE Progress Report No. 25, Two-Year Research Summary – The American West, Greece, and Germany*, Journal of Orgonomy, Vol. 25, No. 2, Orgonomic Publications Inc, New York, November 1991. Also see DeMeo, J.: *Research Progress Report. Cloudbusting in the High Sierras, Central California*, Pulse of the Planet #3, Ashland, Oregon (USA), 1991.

30. The data was drawn from maps supplied by the *National Weather Service*, which show the percentage of rainfall relative to the month of April 1990 for the western parts of the USA.

31. DeMeo, J.: *CORE Progress Report No. 25, Two-Year Research Summary – The American West, Greece, and Germany*, Journal of Orgonomy, Vol. 25, No. 2, Orgonomic Publications Inc, New York, November 1991. Also see DeMeo, J.: *Research Progress Report. Cloudbusting in Northeast Montana*, Pulse of the Planet #3, Ashland, Oregon (USA), 1991.

32. DeMeo, J.: *CORE Progress Report No. 26, California Drought of 1990-1991. Part II*, Journal of Orgonomy, Vol. 26, No. 1, Orgonomic Publications Inc, New York, May 1992.

33. Data supplied by *Climatological Data,* USDOC, NOAA, related to 380 NWS weather stations, 1990-1991.

34. DeMeo, J.: *CORE Progress Report No. 25, Two-Year Research Summary – The American West, Greece, and Germany*, Journal of Orgonomy, Vol. 25, No. 2, Orgonomic Publications Inc, New York, November 1991. Also see DeMeo, J.: *Research Project Report. Greece 1990*, Pulse of the Planet #3, Natural Energy Works, Ashland, Oregon (USA), 1991.
35. DeMeo, J.: *CORE Progress Report No. 30, The Desert Greening Project in Israel 1991-1992*, Journal of Orgonomy, Vol. 26, No. 2, Orgonomic Publications Inc, New York, autumn/winter 1992. Also see DeMeo, J.: *OROP Israel 1991-1992. A Cloudbusting Experiment to Restore Wintertime Rains to Israel and the Eastern Mediterranean During an Extended Period of Drought*, Pulse of the Planet #4, Natural Energy Works, Ashland, Oregon (USA), 1993. Also see DeMeo, J.: *Research Report & Observations. OROP Israel 1991-1992*, Pulse of the Planet #4, Natural Energy Works, Ashland, Oregon (USA), 1993.
36. Rosen, R.: *Report on Cloud Busting Operations and Rain Fall*, unpublished data, 1970; also see Rosen, R.: *News and Comment: Weather Control in Israel*, Creative Process, III (1), 1963.
37. Gassner, M.: *Orgonomy in Israel: Yesterday and Today*, Offshoots of Orgonomy, spring 1985.
38. Greenfield, J.: *Between Orgonomy and Jewishness*, Energy and Character, 7(3):58-59, 1976.
39. Blasband, R.: *Summary Report of Orgonomic Weather Control Operations in Israel*, 1983, unpublished data, in DeMeo, J.: *CORE Report No.30, The Desert Greening Project in Israel 1991-1992*, Journal of Orgonomy, Vol. 26, No. 2, Orgonomic Publications Inc, New York, autumn/winter 1992.
40. DeMeo, J.: *OROP Israel 1991-1992. A Cloudbusting Experiment to Restore Wintertime Rains to Israel and the Eastern Mediterranean During an Extended Period of Drought*, Pulse of the Planet #4, Natural Energy Works, Ashland, Oregon (USA), 1993.
41. Rudge, D.: *Flying in the Eye of a Storm*, The Jerusalem Post, 7 February 1992.
42. *Jerusalem Post*, 27 June 1992.
43. Data supplied by *Agroclimatological Summary*, Bet Dagan Climate Center, Israel, 1991-1992.
44. DeMeo, J.: *Research Reports & Observations. OROP Namibia 1992-1993*, Pulse of the Planet #4, Natural Energy Works, Ashland, Oregon (USA), 1993.
45. Senf, B.: *Die Wiederentdeckung des Lebendigen. Erforschung der Lebensenergie durch Reich, Schauberger, Lakhovsky, u.a.*, Omega Verlag, Frankfurt, 1996.
46. DeMeo, J.: *Green Sea Eritrea: A 5-year Desert-Greening CORE Project in the SE African-Sahel*, Pulse of the Planet #5, Natural Energy Works, Ashland, Oregon (USA), 2002.
47. The data was taken from: *Wettest Rainy Season in 30 Years Across African Sahel*, Special Climate Summary 94/2, Climate Analysis Center, USDOC/NOAA, NWS/NMC, Camp Springs (USA), October 1994.
48. The data was supplied by the Eritrean Ministry of Agriculture in *Daily Rainfall Record of Selected Stations*, 1995. This data was not available in 1994.
49. The data was supplied by the Eritrean Ministry of Agriculture in *Daily Rainfall Record of Selected Stations*, 1996.
50. The data was supplied by the Eritrean Ministry of Agriculture in *Daily Rainfall Record of Selected Stations*, 1997.
51. *El Niño* is an anomaly and an exceptional heating of the ocean's surface water in the eastern-equatorial area of the Pacific. The phenomenon occurs irregularly and happens every 3 to 7 years and can last up to 24 months.
52. The data was supplied by the Eritrean Ministry of Agriculture in *Daily Rainfall Record of Selected Stations*, 1998.
53. Witze, A.: *Lakes Bring New Chance for Life in Sahara*, Dallas Morning News, 5 December 2000.
54. See NASA website at the following address http://visibleearth.nasa.gov/cgi-bin/viewrecord?6056. Also see DeMeo, J.: *Green Sea Eritrea: A 5-year Desert-Greening CORE Project in the SE African-Sahel*, Pulse of the Planet #5, Natural Energy Works, Ashland, Oregon (USA), 2002.

Other Cloudbusting Experiments

".... During my lifetime, I have dedicated myself to this struggle of the African people. I have fought against white domination and I have fought against black domination. I have cherished the ideal of a democratic and free society in which all persons live together in harmony and with equal opportunities. It is an ideal which I hope to live for and to achieve. But if needs be, it is an ideal for which I am prepared to die ...."

Nelson Mandela

The experiences of *cloudbusting* around the world are quite limited, and few groups exist that do research in this direction, apart from the ones already mentioned. Among the most important are those coordinated by Bernardo Zanini and Mirko Kulig, who both work in south central Europe, and Trevor J. Constable, who works mainly in the USA and Asia. In addition to validating their work in an objective and scientific way, these researchers have tried to develop new solutions for a safer and more secure *cloudbusting*.

Bernardo Zanini

The activities of the Zanini group mainly took place in southern Europe.[1] Zanini gained his experiences using both the traditional *cloudbuster* and non-conventional ones. The first orgonomic research of his group was carried out at the start of the 1990s with the development of a method to detect the presence of orgone energy in a qualitative way. Using a bismuth-based solution, they saw that it was possible to determine the presence of orgone in the environment. In Photo 10, two test tubes containing the same solution are shown. The right tube had been kept inside an orgone energy accumulator for a certain amount of time, while the left one had been placed in a control box. As one can clearly see in the picture, the liquid in the test tube from the accumulator has a darker color.

The first *cloudbusting* experiments were carried out with an non-conventional type of instrument. Afterwards, they began experimenting with a traditional *cloudbuster*, made of five steel tubes on a wooden structure fixed to the ground. The tubes were 1.5 meters in length, had a diameter of 3 centimeters and could be connected to extensions for a maximum length of 3 meters. They were grounded into a container of circulating water. The first tests with this tool were undertaken towards the end of the 1990s. Zanini recounts the following about the first attempts at dissolving clouds:

"The day we selected was a very good one for the intervention because the sky was really cloudy with black and white clouds, there was little light and DOR was present in the air....

We turned on the pump and pointed the tubes of the cloudbuster at the clouds in front of us, and incredibly, within an hour we were able to open a pathway in the clouds and once again could see the blue sky. Even nature was re-awakening as if by magic, the blackbirds and sparrows started to fly and tweet again, and the trees once more raised their branches that had hung tiredly.

During the following experiments, these phenomena were repeated, thus giving raise to the conviction that during and after the operations with the cloudbuster, an orgonic radiation formed and remained in the atmosphere for a certain period of time that we called the cloudbusting effect, probably photographable with an infrared film.... "[2]

Photo 11 reports some moments of one of the numerous experiences aimed at dissolving clouds, carried out in September 1998. The pictures, taken at intervals of 5 minutes, detail the rapid movement of the clouds and the final dissolution in the zone surrounding the area where the tubes had been pointed. Photo 12 shows the device built and used by Zanini and Carmelodi during the experiments.

The weather forecast for that day predicted cloudy weather for at least another 5 consecutive days, with a tendency to rain in Lombardy and in the neighboring areas. The day was cloudy with many black and white clouds, and without wind. There was little brightness in the first hours of the morning, and nature itself seemed to stand still. This phenomenon was probably due to the presence of an elevated quantity of DOR in the atmosphere. Zanini describes that experience as follows:

"We started the test at 15:00 hours. Our cloudbuster, called Jew's Harp, was placed with the tubes pointed at 55 degrees towards the west where the sun sets.

The tubes' extensions had not yet been inserted; we waited about half an hour without seeing any changes. The cloud masses passed slowly in front of our position, without alterations. After this, we inserted into each of the tubes of the cloudbuster 5 aluminium rods of about one meter in length, wrapped in newspaper in order to create a difference of potential. We then lengthened the tubes to three meters with the appropriate extensions.

The clouds started to slowly dissolve in the area of the sky where the tubes were pointed, until they disappeared. After about one hour, the clouds had completely dissipated.

After inserting the tubes' extensions, the atmosphere became brighter. This was easily verifiable through a manual exposure meter. Nature seemed to re-awaken. At the end of the operation, we had a strong headache with a sense of dizziness, nausea and intense thirst. "[2]

Experiments were also carried out to remove winter fog. These were performed on the foggiest days with a humidity value of around 100%. Before each intervention, the weather forecast was always consulted on the internet and teletext so as to ensure not to happen upon a day when a clearing up had been forecasted. The procedure used included very gentle clockwise circular movements of the *cloudbuster*. In the majority of cases, a positive result was achieved with the removal of the fog within a radius of 50 kilometers.

Contrary to the interventions for the dispersion of the clouds, which were intended to bring clear weather, those aimed at creating the clouds were much more difficult to carry out, especially in the summer months when the atmosphere is heavy with the so-called *cloak* of very high temperatures and humidity. In these cases, it was necessary to carry out operations intending to influence a wide radius, which could cover southeastern Europe. The operations were always undertaken at intervals of one week, so as to not create climatic disturbances. In addition, the interventions were never longer than two

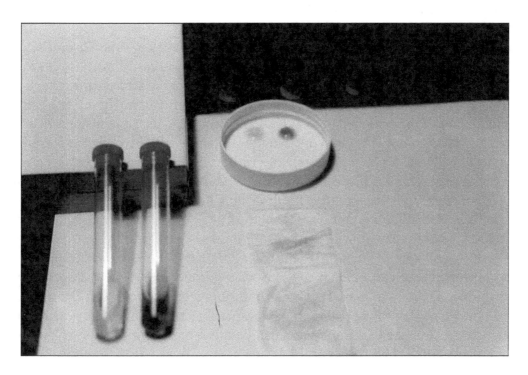

**Photo 10 – Test to Determine the Concentration of Orgone Energy
(Courtesy of Bernardo Zanini)**

**Photo 11 – Cloudbusting For Cloud Dissipation, September 1998
(Courtesy of Bernardo Zanini)**

hours each, and the weather forecasts were always checked in order to verify that disturbances had not set in that could have modified the atmospheric conditions.

A big limitation in the operation of the *cloudbuster* is the necessity of running water. Very often the availability of running water is very scarce, and this limits its use to areas where it is easy to find running water. To make it possible to also use it in areas where running water is difficult to find, Zanini developed a few prototypes of a *cloudbuster* without circulating water. These are equipped with an orgone accumulator and a fan connected to a car battery. With these instruments, it is believed possible to both send and absorb orgone energy into or from the atmosphere, depending upon rotating direction of the fan connected to the orgone accumulator. In addition, the fan motor is equipped with a potentiometer which makes it possible to vary the speed, and hence the orgone energy supplied by the accumulator in accordance with the operative needs. Using these devices, it is also possible to carry out operations for more consecutive days, and Zanini believes it to be less dangerous than the running water *cloudbuster*. Photo 13 shows a version of the fan *cloudbuster*.

Zanini has always paid a lot of attention to the safety of the people who work near the *cloudbuster*. In fact, even a homeopathic doctor belongs to the working group who treats the energetic overcharge of the organism during the interventions. This is one of the biggest problems the Zanini group had, since it was observed that very often the nearness of devices such as the orgone energy accumulator or the *cloudbuster*, can create illnesses in the operators that could also cause irreparable physical damage, if the energetic overcharge of the body is not properly discharged. Based on his experiences, Zanini stated the following:

"... Working in a laboratory of about 15 square meters, our organisms were so charged with orgone energy that we continuously suffered disorders such as nausea, headaches, insomnia, irritability, which had to be cured repeatedly by a homeopathic doctor friend of mine ..."[2]

He advised great caution in the use of these instruments and not to keep orgone devices in the apartment or close to the area where you live, otherwise you could put your health in danger. During one of the operations that he carried out, he experienced an overcharge of energy that was so high and immediate that it was difficult to continue with the operation. Zanini writes:

"... Almost immediately we felt sick, with a sensation of pressure in the head, a metallic taste in the mouth, and a strong desire to drink. It was clear that the operation was contaminating us. We found a way to approximately measure the energy around and within us. We used a needle tester [galvanometer] that functioned without a battery on the micro-Ampere scale, equipped with a handle about 30 centimeters long and isolated in the middle with a winding of copper and zinc. It was sufficient to hold it to have a simultaneous reading.

Before the test the needle slightly moved, but during the experiment we had to change the scale, a clear sign that the energy within us was increasing ...

The most surprising thing happened a few days later. We became so hypersensitive working around the orgone accumulator and the cloudbuster that we were even able to feel the photocell at the entrance to a big department store... Several days later, we observed that the clouds passing over the site where the operation took place were dissolving until they disappeared. This phenomenon lasted for about 20 days. I felt a general sense of illness up to approx. 20 kilometers distance from the test site, until it finally lessened and vanished. I re-read all the material on cloudbusting available to me

**Photo 12 – Running Water Cloudbuster,
Operated by Carmelodi Doz
(Courtesy of Bernardo Zanini)**

**Photo 13 – Fan Cloudbuster
(Without Running Water),
Operated by Bernardo Zanini (Courtesy
of Bernardo Zanini)**

in Italian, to no avail. I was unable to find an explanation."[2]

Through the modified galvanometer, it is possible to measure an increase in the energy level of the body even when operating at a distance of 50 meters from the *cloudbuster*. Photos 14 and 15 show the modified tester and the way to measure the energy charge of the organism.

A step forward in the understanding of the behavior of a circulating water *cloudbuster* was recently taken by Carmelodi, Maglione and Zanini.[3] In a laboratory test, they identified the presence of a direct current between the upper end of the pipe pointed at the atmosphere and the water table in which the tube was immersed. The measurements were carried out with a galvanometer that had the battery removed. Figure 44 shows the graph of the measurements taken at different flow rates.

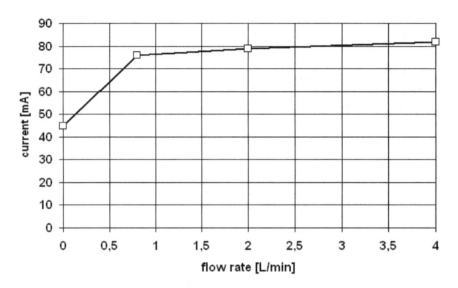

Figure 44: Galvanometer Results (from Carmelodi et al.[3])

As can be seen from the figure, a current also exists under static conditions without the presence of flowing water, and increases with the rise in the water's flow rate.

These results are very similar to those obtained by Volta, who found that between two metallic conductors, separated by an electrolyte, a potential difference is established that puts in motion the electric charges giving rise to a current. It is therefore possible that the working of a *cloudbuster*, when using circulating water, and its dangerous effects on the people that work near to it, could have an ionic origin. Similarly, Nikolaidis[4] observed the variation of an energy field all around a single vertical metal pipe immersed in a steel barrel with running water, that could be compared to excited orgone energy. Zanini concludes his report with a warning about the risks of the use of the *cloudbuster*, both on a personal and on an environmental level:

> "... *The cloudbuster is an instrument that only serves to lend a helping hand to the atmosphere to regenerate itself to the normal conditions of rain and good weather. In addition, it can intervene with positive results against atmospheric pollution caused by vehicle exhausts and smoke from home and office heating* ...
> "*It must be stressed that a cloudbuster is not a toy and that its use in an irresponsible way, aside from creating environmental damage, can cause irreversible harmful effects both to the operator and to people who find themselves in its immediate vicinity. This*

Photo 14 – Galvanometer Modified to Determine the Energetic Charge of an Organism (Courtesy of Bernardo Zanini)

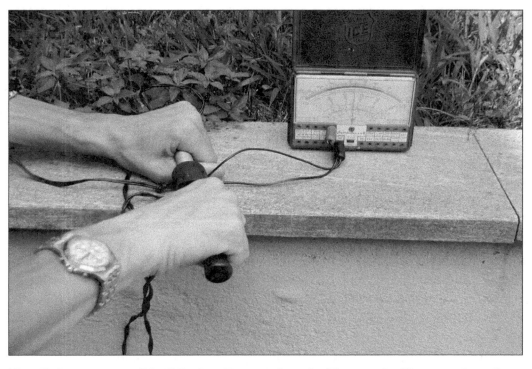

Photo 15 – Galvanometer Modified to Determine the Energetic Charge of an Organism, in Use (Courtesy of Bernardo Zanini)

apparatus therefore requires a great deal of responsibility, a knowledge of the weather, and mastery of the device and the cloudbusting techniques for changing the weather..."[2]

The *cloudbuster* operator has to remain at a safe distance from the energetic field of the working apparatus in order to considerably lower its harmful effects on the health.

Trevor J. Constable

Constable became interested in the field of orgonomy in the late 1950s. The majority of his own research was directed towards stimulating the presence of biocellular energy forms invisible to the naked eye, which he photographed and made visible through the use of particular infrared techniques.[5] Around same time, he started to carry out *cloudbusting* experiments, mostly in the United States and in South East Asia (Photo 16). Among his closest collaborators was Robert McCullough (Photo 17), who was Reich's assistant in the 1950s and who participated in all operations undertaken during that period. The majority of Constable's experiences were collected in a book[6] and a video[7] presenting his philosophy and the majority of the results of his experiments.

From his research activities in the world of the invisible infrared, he proposed new laws on the behavior of Reich's traditional *cloudbuster*. He formulated a totally new concept on its functioning, based on special films he made during its use.[8] Similar experiments were carried out by Boccone[9] in Italy and Cox in England. They obtained the same results with many photographic evidences of dark, even black elliptical and circular forms in the atmosphere. However, Collins[10] reproduced the experiments performed by Constable and his followers in 1994, using the same infrared photographic method but under more controlled experimental conditions. He devised a stereo camera which made identical pairs of photos shot at the same exact time through two different lenses. Collins obtained results similar to those by Constable, but the dark forms did not duplicate on both sides of the stereo pair images of his films, arising criticism about the method and results of Constable, and raising questions about the reliability of his explanations of this still unexplained phenomenon.

Constable argued that the *cloudbuster* worked more to send out energy than being an energy absorber. He believed that during operations, it sent out beams of orgone energy into the atmosphere at a theoretically infinite distance, rather than absorb and discharge the energy into running water, as Reich had established. Constable considered each pipe to be generating a kind of laser-ray of orgone, with the difference that the energy projected into the atmosphere was of a bioenergetic type and not an electromagnetic one. By Constable's theory, the changes in the atmosphere's orgonomic potential occurred due to the beams of orgone energy shot by the pipes. This would increase the energy potential of the atmosphere in the area at which the pipes are directed. Figure 45 schematically illustrates Constable's view of the phenomenon.

It is apparent that this concept finds itself in sharp contrast with the theories proposed by Reich and subsequently confirmed by Blasband and DeMeo. However, Constable supplied his own explanation for the phenomena associated with the use of the *cloudbuster* and for the origin of the atmospheric events connected with it. He claimed that the holes forming in the cloud cover when the pipes are pointed at the sky, were created by the high orgone potential of the beams fired from those pipes. While passing through the clouds, the beams of energy absorbed all the humidity present in their immediate vicinity, creating an area of low humidity and low orgonomic potential. In the same way, he felt this explained the breaking up and dissolution of the clouds. Furthermore, he argued the *cloudbuster* would absorb energy from the area surrounding its position during operation, thus lowering the orgonomic potential. The tendency of an operator standing close to the device to become bioenergetically over charged, he

**Photo 16 – Running Water Cloudbuster *Willy's Wand*, 1970
(Courtesy of Trevor J. Constable)**

**Figure 45:
Constable's Theory**

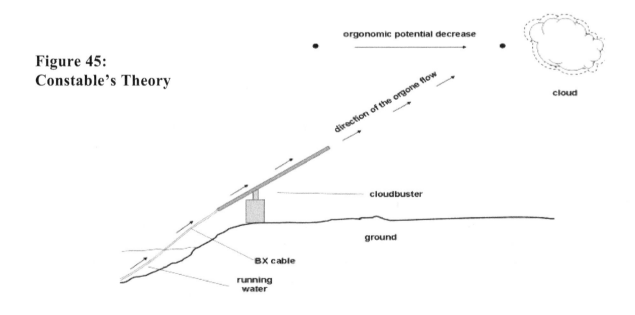

Photo 17 – Running Water Cloudbuster *Magnum 144*, Operated by Robert McCullough During an Operation in the Mojave Desert, Southern California (Courtesy of Trevor J. Constable)

felt confirmed this hypothesis, but only through the mechanism where the surrounding environment possessed a lower orgonomic potential than that of human beings. He claimed this occurred through the absorption of orgone energy by the *cloudbuster* in the area surrounding the device, which lowers the orgonomic potential to the point where orgone energy is more easily absorbed by the organism.

In addition to all these theoretical changes, Constable also devised *cloudbusters* working without circulating water, based upon his experimental observations. He used particular geometric shapes to create instruments that fired energy into the atmosphere.

The reactions to the theories proposed by Constable were rather restrained. Based on his experiences of 35 years of research with the traditional *cloudbuster*, Blasband admitted[11] that it could work in both ways depending on the state of the atmosphere, the manner in which the tubes were immersed into water, and other factors. He observed that sometimes the device seemed to work mostly as a passive conductor of orgone energy from the earth's surface towards the atmosphere. This phenomenon was basically a function of the orgonomic potential of the earth and the atmosphere in that moment. Therefore, a lot of attention needed to be paid to its behavior during use, because if used unwittingly as an emitter of orgone energy, it could amplify the already precarious conditions of the atmosphere instead of reducing them. Blasband was convinced that further research was necessary, particularly in order to develop measuring instruments for orgone energy, and to better understand how the *cloudbuster* functioned.

Among the numerous operations carried out by Constable with the traditional apparatus are those to reduce smog and high temperatures in specific areas of the West Coast. In operation *Kooler*[12], he intervened to reduce the high temperatures and smog levels that had hit a part of Southern California particularly in concentrated urban zones. Consequently, he did not attempt to bring rain or alleviate the drought situation, but to clean the atmosphere of pollutants and cool down certain areas. In any case, the principle Constable based his operations on was always that of the law of the orgonomic potentials, but adapted for the formation of light breezes, necessary to create an air movement in the atmosphere. By pointing the tubes in a certain direction, a light breeze can be created that blows from that direction. Once the breeze has formed, the direction of the tubes can be changed, and within a short period of time, a light breeze blowing in the new direction comes up. Slight winds can develop much easier if the *cloudbuster* is pointed westwards, yet they can be triggered in whatever direction it is pointed. Ola Raknes, a student and follower of Reich, wrote about his experiences at Reich's side on this topic:[13]

> *"... One day Reich was experimenting with the cloudbuster, trying to find out in which direction it would be most profitable to point the apparatus. While he was pointing it in different directions, I happened to notice that the wind, a light breeze over the nearby Dodge Pond, was changing its direction. Not far from the cloudbuster was a weathervane, and I decided to follow its movements. Whenever Reich changed the direction of the cloudbuster, the weathervane would show in a few minutes that the wind had changed correspondingly. I was strongly impressed by this observation ..."* [13]

Constable stated that this phenomenon can be observed very easily in the absence of wind or if only a slight movement of the air exists. In addition, as Raknes testified to, there is an almost immediate correspondence, or at the most of some minutes, between the action of the *cloudbuster* and the formation of a light breeze, especially under stagnant atmospheric conditions. In case there is a fairly consistent wind, a few hours could be necessary to create a breeze that blows along the absorption axis. Among his many activities, Constable performed many of these operations, sometimes even very difficult ones like stopping the strong hot Santa Ana winds which blow in southern California.

Operation *Kooler*[12] was set for mid-September 1971 to eliminate the smog and lower the high temperatures that had hit Southern California, and the Los Angeles area in particular where a maximum of 106° F[14] had been registered. The weather forecast for September 12th had predicted no changes for the following days, a testimony to the stasis condition the atmosphere was in. The operation, aimed at creating a light wind in the area to get rid of the smog, started on September 12th at 18:45. Just about 40 hours after the start of the operation, there was a major drop in temperature in the central Los Angeles area of around 9° F. The reduction was even more significant on the next day, when the temperature fell by 31° F, then settled on a value of 75-76° F. Also a light rain followed the lowering of the temperature, which had not been the intent of the intervention. The operation ended on the 16th of September at midnight. Figure 46 presents the maximum temperature trend registered in the central Los Angeles area during the time of the operation.

As the graph shows, the total reduction of the temperature during the operation was around 31° F. This event proved once again the effectiveness of the interventions, this time aimed at lowering the temperature and reducing the heat and the smog level.

One special case that Constable described in his book *Loom of the Future*[15], was an intervention carried out in the second half of the 1980s to reduce the smog in central California. This highly industrialized area had a high level of atmospheric pollution, evaluated by the number of days when the smog levels rose above the permissible limit.

In the years 1984 to 1986, the area suffered elevated smog levels on the alarming number of 97, 83, and 79 days, respectively. Constable intervened with a long-term plan to reduce the level of smog. With his group of around 5 to 6 people, he performed absorptions intended to create breezes and air flows in different areas of Southern California to get rid of the smog. These were operations *Victor* in 1987,

Breakthru & Checker in 1989, and *Clincher* in 1990. As a result, a smog reduction of 16.6% took place in 1987, in 1989 it was 29.4%, and finally 24% in 1990. The number of days with elevated smog levels dropped from 79 days in 1986 (before operation *Victor*) to 41 days in 1990.

Urban centers such as Pasadena saw a reduction in smog from 33 days in 1987 to 7 days in 1990, the same happened in central Los Angeles, where it went from 8 down to 2 days. The operations were suspended in 1991, and that year there was an increase of 12% (46 days). Commenting on the results of the operations, Constable emphasized the effectiveness of *cloudbusting* and especially the low cost of the interventions, compared to the budget assigned for conventional air-pollution control methods. The cost of combating the smog in Southern California with conventional methods was estimated at around 20 billion dollars in 10 years, and with an average cost of 2 billion dollars per year. The total cost of operation *Clincher*, with 14 operative bases spread throughout all of Southern California, was approximately 55 thousand dollars, a sum which is *less than 1%* of the cost spent by the administration for conventional air pollution controls in just one day.

Figure 47 shows the development of smog levels for central California from 1984 to 1991.

A last case was the *Pincer II* operation[16], planned to bring rain into the city of Los Angeles in July 1986. The month of July in Los Angeles is well known for its statistical lack of rain that dates back to the beginning of the last century. Bringing rain to the city of Los Angeles in July requires that the normal, natural passage of Mexican moisture out of the Gulf of California northwards and northeastwards into Arizona be diverted, with a near-90 degree bend. Thus the normal orgone flow will go about 250 miles northwestwards out of its way. The intervention to bring moisture into the Los Angeles basin was planned to occur through two routes designed like a pincer, so that they converged on the Los Angeles Civic Center.

Photo 18 shows the detailed engineering drawing of the operational plan. The blue flow (marked "C" on the image) is the normal July path of moisture from the Gulf of California that produces the Arizona monsoon season. The red flow ("B"), one of the arms of the pincer, is the engineered diversion of the blue stream, causing the moisture to move into central Southern California. To bend this flow, two *cloudbuster* bases were used located at Fort Z and Banning. The green flow ("A"), the second arm of the pincer, induced moisture to flow northwest from the Gulf of California, to the Coast towards the southerly tip of Los Angeles. Two *cloudbuster* bases were used also in this case: Hatfield Flat (located

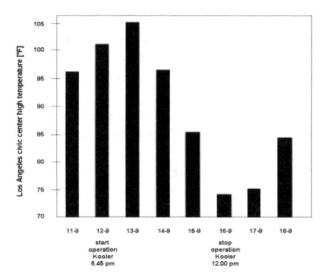

Figure 46: Observed Temperatures During Operation *Kooler*
(from Constable[12])

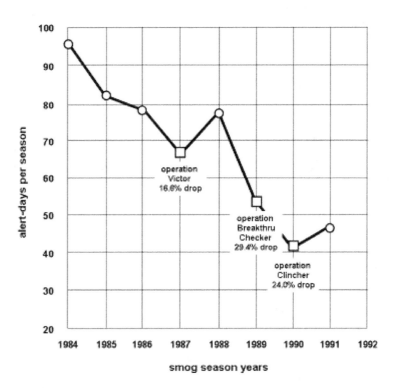

Figure 47: Southern California Smog Alert-Days Per Season, 1984-1991 (from Constable[15])

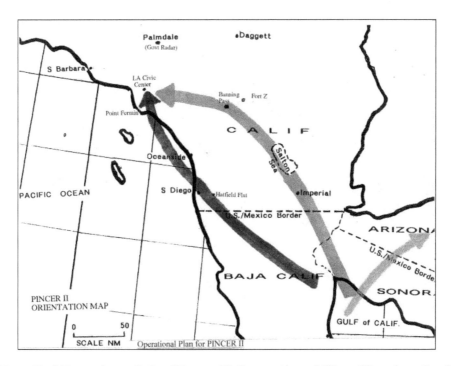

Photo 18 – Detailed Drawing of the *Pincer II* Operational Plan Showing the Engineered Diversion [Green (A) and Red (B) Flows] of the Normal July Path of Moisture [Blue (C) Flow], from the Gulf of Mexico into Arizona (Courtesy of Trevor J. Constable)

Photo 19 – Green Route (Left) Induces Moisture Northward From the Gulf of California to Flow Up to the Southerly Tip of Los Angeles. Red Route (Right) Causes Moisture to Move to L.A. Civic Center Through Central Southern California. (Courtesy of Trevor J. Constable)

to the east of San Diego), and Point Fermin.

The *Pincer II* operation went through three distinct phases. Phase 1 started on the 4th of July 1986. The weather bulletin forecasted hot weather for the weekend, but following the interventions a very cool weekend ensued with rain in San Diego and Pasadena. Phase 2 ended on July 15th with a two and a half hour thunderstorm with lightning in San Pedro bay. This was not forecasted by the weather bulletin.

Phase 3 consisted of channelling the moisture from Mexico on the 22nd of July, via both arms of the pincer. The moisture was diverted along the red and the green routes to the center of Los Angeles. The red flow came first, in the late afternoon and early evening of July 22nd. As a consequence, the Los Angeles basin was full of rain by 19:30. The green flow reached Los Angeles by 1:30 on July 23rd (Photo 19). Los Angeles Civic Center recorded 0.18 inches of rain, making it the wettest July in 100 years. Besides, 72 out of the 100 official weather stations located in Southern California recorded rain in July of 1986. Additionally, rain was produced in National Forest areas, quenching a number of stubborn forest fires. Some official stations recorded 4 inches of rain during the month of July, causing the fires in the National Forest to decline sharply.

Mirko Kulig

The group of Kulig works mainly in south central Europe. The activities of the group are principally devoted to experiments with modified versions of the Reich *cloudbuster*. Their instruments consist mainly of tubes connected to an orgone energy accumulator. The function is claimed to radiate energy into the atmosphere, instead of absorbing it, as taking place in the traditional *cloudbuster*.

Their experiences with the traditional version of the *cloudbuster* were only very few. In one such case, a significant test was undertaken in the Sahel Zone of Africa (Photo 20), in the village of Salmossi in Burkina Faso (Photo 21), in February of 2001. Kulig carried out an attempt with a very rudimentary instrument, built with about twenty iron tubes of 1.5-2 meters in length, vertically fixed in the center of a small water well. The results were interesting. They acquired a partial cloud cover (never before observed during nighttime hours) and an increase in the humidity, which surprised the majority of the people present in the village.

Photo 22 shows the rudimentary *cloudbuster* built and used in this experiment. The tubes of different lengths were driven vertically into the pond of water situated close to the village. Kulig described his experience as follows[17]:

> "*I found myself in the village of Salmossi, in the north of Burkina Faso, on the thin stretch of land that separates the Sahara desert from the African tropical forest and which is called the Sahel.*
>
> *I was the assistant/guide of a group of European sixteen year olds who had decided to "work in the field" in the course of a project against desertification with a method developed by a well-known Italian agronomist. Our work consisted of teaching the Africans, natives of the area, the seeding techniques to adopt on the soil treated with this method, but yet more in helping them to realize that their work could guarantee to the population who live in that area that they could stay in their villages without having to emigrate due to a lack of water. As an observer, I can say that this conventional method is the most efficient and economic that I have seen to initiate fighting against the advancing desert.*
>
> *I had just finished studying Reich's books that dealt with the operations and theories of cloudbusting, therefore for all the time that I was in the village of Salmossi (about 2*

weeks), I carefully observed the environment and the climate to better understand the dynamics that cause an area to become a desert (and that area had a high risk of desertification). The last evening before leaving the village, I couldn't resist the temptation and decided to do a cloudbusting experiment. There were very few means available, but the principal elements were there: the iron tubes that are used for seeding and a well of water of 1 to 2 m². I simply drove the tubes (around 20 – 25), which had a length of 1.5 to 2 m, into the well pointing at the zenith. It was 17:30 in the evening. The first observations were: an increase in the cloud cover, with an area free of clouds exactly overhead. When it became dark (at around 18:30), we noticed that we could not observe the stars and the constellations as we had on other evenings: the sky was almost totally covered. It must be noted that, even if the sky was sometimes slightly cloudy during the day, in the evening we were always able to make our observations of the stars. That evening however, the clouds remained even with the coming of the darkness. That last evening, there was a large gathering of people (locals and others) surrounding our camp, waiting for the farewell party. They therefore didn't pay much attention to the tubes that had been placed in the well, because I had already informed the teenagers and the other accompanying people of my studies on Reich's theories. A group of teenagers was particularly interested in the discussion and helped me to make our primitive cloudbuster. As soon as the teenagers saw that the sky was cloudy they became excited, which I was barely able to control, indicating the seriousness of the experiment and the possible consequences.

The device remained in place until two in the morning, and during all that time I regularly poured water into the tubes (remembering that according to Reich, the power of a cloudbuster is increased if the water is moving).

The following observations were made: many people of the group said the air was easier to breathe, as if the relative humidity had increased. Unfortunately I didn't have a hygrometer with me, so we could not confirm this fact, but I want to emphasize that during that time, the relative humidity was certainly no higher than 20% (based on the fact that a pair of jeans that had just been washed took only an hour and a half to completely dry), therefore an increase in the humidity was immediately perceived as a relief on a physical level. During this experiment, I observed that cloud fronts coming at regular intervals from the southeast, when arriving in our vicinity, lost consistency (this observation was possible based on the visibility of the stars). At a certain point, the sky was completely covered for about 30 minutes, and this was the most surprising thing. In addition, we discussed this fact on the next day with the agronomist who stayed in the nearby town of Gorom-Gorom (about 20 km away), and it was revealed that the sky had been cloudy in the area the evening before, the agronomist was astounded. This was an indication of how much influence an operation of this kind can have.

My most demanding question after the experiment was: how was it possible to obtain those results with so few tubes and such a small well of water? It then hit me that there were other wells surround the village (9 in total). Is it possible that the water in the small well in which I had placed the tubes, seeping through the ground, could create a connection with the ground water table? I thought that this was the most logical explanation to justify such notable effects with such basic tools. Many objections can be made as to the scientific nature of this experiment and the related observations, and my aim was not to do an experiment to prove certain theories, but the one thing I can guarantee is that all people present have no doubt that the phenomena we observed were

Photo 20 – African Sahel (Burkina Faso, West Africa)
(Courtesy of Mirko Kulig)

Photo 21 – Village of Salmossi (Burkina Faso, West Africa)
(Courtesy of Mirko Kulig)

**Photo 22 – Cloudbuster Used in the Salmossi Experiment.
Reproduced in Color on Rear Cover. (Courtesy of Mirko Kulig)**

due to the operation. The following morning, we had to leave to return to the capital, and as we got up at around six, we noticed that there were still clouds in the sky (high cumulous) that were slowly dissolving with the sunrise."[17]

In my view, it is important to point out that the subjective perception of Kulig and the other people present at the test, can somewhat confirm its positive effect. Also the fact that many people were surprised by the noticeable changes in the weather and in some of the atmospheric parameters of that area is an indicator assuring the effectiveness of the device used. It is even more astounding if one considers the simplicity of the materials used: a few tubes immersed in a well of water.

Other Experiments

Other people have conducted experiments over the last years. Very often, they only became curious by reading the works of Reich or one of his successors, and so they personally constructed and developed simple instruments, to test the functionality and reliability of the methods employed.

As the atmosphere and the orgone energy envelope surrounding the globe is finite, and the absorption influences of the instruments do not reach beyond a certain distance, one operator who intends to do *cloudbusting* cannot be aware of what other operators are doing in the same area and in the same time period. It is therefore not surprising if the results obtained are not satisfying or even opposite of what is expected. It is very important to always exchange information on your plans and activities with other operators, so the operations in the same area can be meticulously coordinated to avoid overlapping on the same day. This also prevents creating catastrophic effects both on the weather and the area in question due to excessive absorptions.

With this goal in mind, a group of American scientists, physicians and naturalists formed the *Core Network* in the early 1990s. This is a professional association to provide for better coordination of experiments and operations with the standard *cloudbuster*, undertaken at different locations primarily in the USA. Such work had been on-going by the members since 1960s with a less formal structure. Today, however, the *Core Network* provides for inter-group discussions of weather problems and phenomena, notifications of operations as undertaken by its various members and apprentices, with occasional meetings and publications. They perform interventions to solve problems of drought and deserts both in the USA and worldwide, and some members also offer seminars and courses for professional training.[18]

In addition, there are a few forums on the internet where issues of *cloudbusting* and orgonomy in general can be discussed. Among the topics discussed in these forums, the safety in the use of the *cloudbuster* is one of the most important issues dealt with. One experimenter advised that great care must be taken with these tools and in particular said that *any metallic part of the cloudbuster must never be touched with a bare hand,* as otherwise a large quantity of orgone energy is transferred to the operator, creating irreparable damage to their health. He discussed this point in a message on the site:

> *"...Remember that any metal connectors you attach to the hose end will conduct OR (orgone) to YOU if you are holding it...... I haven't mentioned it before, but I have an heart condition. Every time I draw, it puts me into an irregular heartbeat. Yesterday was no exception ..... As I directed the pipes, I did the forbidden, and held the insulated cable(s) in my right hand..... I was gently sweeping and felt a gentle vibration through the cable on the fingers of my right hand. I noticed it was variable and felt it the strongest at the highest concentration of energy. I felt like I found a vein in the atmosphere. I'll bet it would have shown up on an oscilloscope. My health is suffering from it today ..... A CAUTION TO EVERYONE WHO UNDERTAKES THIS WORK!!!"*

In his intervention, it was essentially confirmed that the physical contact with the metal parts can alter the heartbeat and therefore create consequences that are also serious for the circulatory system.

It is of fundamental importance when moving the metallic parts of the *cloudbuster*, not to come into direct contact with them. It is strongly advised, even by Reich himself who was the first to notice the danger of the discharge of this energy on the human body, to use strong cotton or rubber gloves. In all of his operations, DeMeo[19] moves the metallic parts of the device by remote controls, either using electrically-operated motors or long pull-ropes, while standing at distance of 7 to 10 meters from the instrument. DeMeo also wrote a separate paper warning of both the personal and social-environmental dangers when the *cloudbuster* is used by untrained individuals, or for unserious reasons.[20]

One experimenter wrote about the real dangers of modifying the atmospheric processes to their own advantage rather than in favor of the natural atmospheric processes:

"...Where any assisting in increasing or decreasing the potential for rain has taken place, the energy flow and its effects on mass (moisture) in the atmosphere ARE altered. How long the residual effect lasts, I do not know, but it appears, from my observation, to remain for quite some time. The atmosphere is extremely complex in its composition, and I believe it remembers these alterations . In areas where extensive CORE work has taken place (cloudbusting) to increase the potential for rain, the atmosphere has not forgotten. If you view the biosphere as a living organism with a circulatory system, with a liquid content (water), and you stimulate it (cause it to perspire) or to relax, patterns of behavior are remembered so it's prepared for the change the next time. It's a natural, defensive action, sort of a self-preservation process. Again, how long this lasts, I do not know, but from what I see, the effects are still there...."

And finally, as author, my own thoughts:

"...Anyone thinking they can control such a diverse, complex and integrated living system as the weather, is fooling themselves. True, it can be influenced and assisted, but I believe it is beyond man's capacity to control."

**"Love, work and knowledge are the wellsprings of our life.
They should also govern it."**

Wilhelm Reich

Chapter 7 Citations

1. Personal communication with the author.
2. Zanini, D.: *L'energia Orgonica e il Cloudbuster*, Altra Scienza, Year 2, No. 8, September/October 2001, Rome, Italy.
3. Carmelodi, D., Maglione, R., Zanini, D.: *Lab Test on the Functioning of a Running Water Cloudbuster*, Altra Scienza, Year 5, No. 27, October 2004, Rome, Italy.
4. Nikolaidis, N.: *Orgone Energy Field Observations Using the Dowsing Rods*, in *Heretic's Notebook*, Pulse of the Planet #5, p. 163-167, Natural Energy Works, Ashland, Oregon (USA), 2002.
5. Constable, T.J.: *The Cosmic Pulse of Life*, Borderland Sciences Research Foundation, Garberville, California, 1990.
6. Constable, T.J.: *Loom of the Future. The Weather Engineering Work of Trevor James Constable. An Interview Conducted by Thomas J. Brown*, Borderland Sciences Research Foundation, Garberville, California, 1994.
7. Constable, T.J.: *Etheric Weather Engineering. On the High Seas*, VHS Video, Borderland Sciences Research Foundation, Garberville, California.
8. Constable, T.J.: *Orgone Energy Weather Engineering Through the Cloudbuster*, in Future Science, edited by White & Krippner, Anchor Books, New York, 1977.
9. Boccone, L.: *UFO, La realtà Nascosta*, Edizioni Ivaldi Editore, 1980, Genoa, Italy.
10. Collins, A.: *Alien Energy, UFOs, Ritual Landscapes and the Human Mind*, Eagle Wing Books, 1994, Memphis, USA.
11. Blasband, R.A.: *Orgonomic Weather Control. An Overview*, Journal of Orgonomy, Vol. 26, No. 1, Orgonomic Publications Inc, New York, Spring/Summer 1992.
12. Constable, T.J.: *Operation "Kooler". Conquest of a Southern California Heat Wave*, Journal of Orgonomy, Vol. 6, No. 1, Orgonomic Publications Inc, New York, May 1972.
13. Raknes, O.: *Wilhelm Reich and Orgonomy*, St. Martin's Press, New York, 1970.
14. One degree Fahrenheit corresponds to 9/5 of a degree Celsius + 32.
15. Constable, T.J.: *Loom of the Future. The Weather Engineering Work of Trevor James Constable. An Interview Conducted by Thomas J. Brown*, Borderland Sciences Research Foundation, Garberville, California, 1994.
16. Constable, T.J.: *Operation Pincer II. July Rain Engineering, Los Angeles 1986*, The Journal of Borderland Research, Garberville, California, Jan/Feb 1987.
17. Kulig, M.: *Un Esperimento di Cloudbusting nel Sahel*, Altra Scienza, Year 3, No. 15, November/December 2002, Rome, Italy.
18. More info on the Core Network can be found on this website: http://www.cloudbusting.org
19. Albini, C.: *Creazione e Castigo. La Grande Congiura Contro Reich*, Tre Editori Editors, Rome, 1997.
20. DeMeo, J.: *So You Want to Build a Cloudbuster*, Orgone Biophysical Research Lab, 1986.
    http://www.orgonelab.org/sobuildaclb.htm   also see: http://www.orgonelab.org/chemtrails.htm

Healing of Atmospheres

Glossary

Aridity. The results of an area subjected to a prolonged drought. It manifests itself at medium and long-term through the reduction in the frequency and quantity of the rainfall (less than 250 mm per year).

Artificial seeding. See *cloudseeding*.

Bioenergy. A living organism's energy that allows it to function.

Bions. Vesicles of energy that represent the transitional phase between non-living and living matter. They are constantly formed in nature through a disintegration process of inorganic and organic substances, a process that can be experimentally reproduced. They are charged with orgone energy, which is life energy, and can transform into bacteria, or aggregate into clumps to form protozoa.

Biophysics. The science that studies the physical aspects of biological processes and in particular the state of the charge and the movement of the organism's energy.

Biopsychiatry. Psychiatry from an energetic point of view.

Block. Contraction of the organism or atmospheric stagnation that impedes the energy's excitation or free flow.

Character analysis. Originally developed from the traditional psychoanalytical technique of the symptom analysis, which included character attitudes bound with the symptoms. The discovery of the muscular armor made it necessary to develop a new technique that was called vegetotherapy. Following this, the discovery of the organism's orgone energy (bioenergy) and the possibility of having a concentration of orgone energy available through an accumulator, made a further development of the character-analytic vegetotherapy into an orgone biophysical therapy necessary.

Character armor. The sum of the typical character attitudes that an individual develops to block their emotional excitation, and that can be expressed in the rigidity of the body, and in the lack of emotional contact. It is functionally identical to muscular armor.

Cloudbuster. The apparatus used in *cloudbusting* to restore self-regulation of the weather.

Cloudbusting. A group of methods and techniques to re-establish the natural flow of orgone energy in the atmosphere, with its natural pulsating cycles of charging (sun) and discharging (rain).

Cloudseeding. Method for the production of rain based on artificial seeding of clouds, which generally uses dry ice or silver iodide.

CORE. Acronym for Cosmic Orgone Engineering. All the technologies and procedures that have the objective of using of the primordial cosmic orgone energy of the atmosphere and the cosmos.

Cosmic Orgonic Engineering. See CORE.

Desertification. The environmental consequence of drought and aridity that hits a particular area. It is characterized by a scarcity of rainfall, by a continuous erosive process of the wind, and by the formation of sandy surfaces. It is currently expanding on our planet at a speed of 60,000 km^2 per year.

DOR. Acronym for Deadly Orgone. State derived from the energetic reaction of the primordial orgone energy especially to x-rays, electromagnetism, and radioactivity. It is dull black, immobile, toxic, and hungry for oxygen, water, and fresh orgone. Under certain conditions, it can be present both in organisms and in the atmosphere.

Dor-buster. A tool used to treat an organism's orgone energy blocks. Its functioning principle is very similar to that of the *cloudbuster*.

Drought. The result of a continued lack of rain. It is considered serious when agricultural production in a certain area is reduced by 10%, and catastrophic when it falls by 30%. A prolonged drought can cause aridity and possibly the desertification of a certain area.

Emotional plague. The neurotic character that behaves with a destructive action on the social scene, actively working to block the natural functions of the living.

Energetic charge. Amount of orgone energy present in a system.

Equatorial Orgone Stream. Flow of orgone energy that moves in the Earth's atmosphere from west to east.

Galactic Orgone Stream. Flow of orgone energy from beyond the Earth's gravitational field, that moves in the atmosphere from southwest to northeast. It forms a 62° angle with the equatorial plane.

Muscular armor. The sum of the muscular attitudes (chronic muscular spasms) that an individual develops to block the breakthrough of emotions and organ sensations, in particular anxiety against sexual excitation.

Oranur. Orgone energy in an excited state caused by x-rays, electromagnetism, and nuclear energy. The state that precedes the formation of DOR.

Orgasm. The unitary involuntary spasm of all organisms at the climax of the sexual act. Because of its involuntary character and by the general fear of the orgasm, this reflex is blocked in most human beings, especially in civilizations that suppress the genitality of children and adolescents.

Orgastic potency. The capacity of complete abandonment to the involuntary spasms of the organism with the complete discharge of excitation at the climax of the sexual act. It is always absent in neurotic individuals. It presupposes the absence of a pathologic character and muscular armor.

Orgone energy. A mass-free energy in the pre-atomic state, universally present and visually, thermically, and electroscopically demonstrable, as well as by using a Geiger-Muller counter. In living organisms, it is present as bioenergy or life energy. Discovered by Wilhelm Reich between 1936 and 1940.

Orgone physics. A group of laws that regulate the behavior of orgone energy.

Orgonomic. Adjective related to orgonomy.

Orgonomy. The natural science that studies orgone energy functions in nature.

Orgonotic. Quality regarding the capacity of the body or of a condition to hold orgone.

Orgonotic charge. See energetic charge.

Primordial cosmic energy. See orgone energy.

Sex-economy. The branch of orgonomy that deals with the economy of the biological energy (orgone) within the organism.

Stasis. Damming up of orgone energy in an organism or in the atmosphere.

Superimposition. The convergence or the attraction of two systems or streams of orgone energy. It is a common functioning principle in nature, present for example in the creation of matter, in the motion of energy-streams which form hurricanes, and in the sexual embrace between two living organisms.

Vegetotherapy. The therapeutic sexual-economical technique. The aim is to liberate the blocked vegetative energy, so as to give the patient their vegetative mobility back.

Healing of Atmospheres

About the Author

Roberto Maglione has been working in the field of petroleum exploration for numerous years where he covered different positions both in Italy and abroad. Currently he is working for an international cardiovascular bioengineering company as responsible of the Supplier Quality and Auditing Activities. He is also a consultant on drilling fluids hydraulics. Maglione has been Visiting Professor at the Polytechnic of Turin in the period 1999-2001. He is the author of two scientific books, on the Rheology and Hydraulics of Drilling Fluids (1998 and 1999) and of more than 90 technical articles.

Maglione has been studying Reich's theories since the early 1990s. He wrote the following books *Wilhelm Reich e la modificazione del clima* (2004), and *The healing of atmospheres* (2007); and co-authored with Nicola Glielmi *Argomenti Reichiani* (2007), *Wilhelm Reich* (2009), *La distruttività post-encefalitica ed il farabuttismo dialettico nelle strategie di comunicazione* (2009), and *L'arca di Mosè e l'accumulatore di Reich. Una identità oltre i confini del tempo* (2010). He contributed also to the book edited by DeMarchi and Valenzi *Wilhelm Reich. Una straordinaria avventura scientifica e umana* (2007).

Maglione holds a MS degree in mining engineering from the Polytechnic of Turin, Italy, and is a member of numerous scientific associations.

For more books and information on
Wilhelm Reich's discoveries and the subject of
orgone energy research, review the offerings on-line at:

Natural Energy Works
Ashland, Oregon, USA

http://www.naturalenergyworks.net

Printed in the USA
CPSIA information can be obtained
at www.ICGtesting.com
LVHW080623060524
779124LV00008B/306